T0350164

The Flowering of Ecology

Emergence of Natural History

VOLUME 3

The titles published in this series are listed at *brill.com/enh*

The Flowering of Ecology

Maria Sibylla Merian's Caterpillar Book

By

Kay Etheridge

*M. S. Merian's original texts translated
from the German*

By

Michael Ritterson

BRILL

LEIDEN | BOSTON

Cover illustration: *Dicallomera fascelina* (dark tussock moth) and *Taraxacum officinale* (dandelion). Plate 8 from Merian, 1679. *Raupen.* NEV Library/Naturalis, Leiden.

Library of Congress Cataloging-in-Publication Data

Names: Etheridge, A. Kay, 1954– author.
Title: The flowering of ecology : Maria Sibylla Merian's caterpillar book /
   by Kay Etheridge ; M.S. Merian's original texts translated from the
   German by Michael Ritterson.
Description: Leiden ; Boston : Brill, 2021. | Series: Emergence of natural
   history, 2452–3283 ; volume 3 | Includes bibliographical references and
   index.
Identifiers: LCCN 2020036154 | ISBN 9789004284791 (hardback) | ISBN
   9789004284807 (ebook)
Subjects: LCSH: Merian, Maria Sibylla, 1647–1717. Raupen wunderbare
   Verwandelung und sonderbare Blumen-Nahrung. | Caterpillars. |
   Butterflies—Metamorphosis. | Moths—Metamorphosis. | Biological
   illustration.
Classification: LCC QL542 .E84 2021 | DDC 595.7813/92—dc23
LC record available at https://lccn.loc.gov/2020036154

Typeface for the Latin, Greek, and Cyrillic scripts: "Brill". See and download: brill.com/brill-typeface.

ISSN 2452-3283
ISBN 978-90-04-28479-1 (hardback)
ISBN 978-90-04-28480-7 (e-book)

For my parents, Sara and Ray Etheridge

Contents

Foreword

Brian Ogilvie

Maria Sibylla Merian was one of the most remarkable observers of insects of the later seventeenth century, a period when many remarkable individuals turned their attention to those "lesser living creatures."[1] Highly respected by her eighteenth-century successors, including Carl Linnaeus, her star faded in the nineteenth century as entomology became increasingly professionalized, and as inaccurate reproductions of her copperplate engravings circulated far more widely than her original drawings and the books printed during her lifetime.[2] In this volume, Kay Etheridge demonstrates amply that Merian deserves to be recognized, as her contemporaries understood, as one of the pioneers of the study of insects, and in particular, as an innovator in the study of relations between insects and their host plants.

Scholars have engaged in recovering Merian's life and her artistic and scientific work for close to a century. This process of recovery has tended to emphasize two aspects: her gender and her travel to Suriname from 1699 to 1701, a journey that resulted in her 1705 folio publication *Metamorphosis Insectorum Surinamensium* (Metamorphosis of the Insects of Suriname). The present volume makes a significant advance to understanding Merian and her contribution to natural history by drawing our attention instead to her first publication on insects, the 1679 quarto volume titled *Der Raupen Wunderbare Verwandelung und Sonderbare Blumen-nahrung* (The Wondrous Transformation and Particular Food Plants of Caterpillars), and to the years of meticulous observation that lay behind it. Though *Metamorphosis* is a more visually impressive book, especially copies with hand-colored engravings, the 1679 caterpillar book reveals much more about Merian's sustained engagement with the natural world and the methods she developed to investigate it.

It was not unusual for artists in the seventeenth century to turn their attention to insects. They appeared in medieval miniatures, in the drawings, watercolors and oil paintings of Renaissance artists and Dutch still life painters, and in widely circulated woodcuts and engravings. When Maria Sibylla was five or six years old, her older half-brothers Matthaeus the Younger and Caspar were

---

1 The phrase is a loose translation of *minima animalia*, the term used to gloss *insecta* in the title of Thomas Moffett's 1634 *Theater of Insects*. "Smallest living creatures" would be more accurate. For a brief overview of early modern studies of insects, see chapter 1, below.
2 See below, chapter 4, p. 119ff.

preparing the engraved illustrations for the Polish naturalist John Johnstone's *Historiae Naturalis de Insectis Libri III* (Natural History of Insects), which the family firm published in 1653.[3] Most of the illustrations were copied from woodcuts in earlier insect books by Ulisse Aldrovandi and Thomas Moffett. But even if Matthaeus and Caspar studied few or no actual insects themselves, Maria Sibylla would have seen hundreds of different kinds in the engraved plates for the work.

What was unusual about Merian was the detailed, painstaking, repeated observations that she made not only of insects' form but also their behavior and transformations from egg to larva, pupa, and imago (adult), and her careful attention to the food plants that they preferred or required. Earlier, humanistically educated naturalists had made less detailed studies of insects, and the Dutch landscape artist Johannes Goedaert had started observing insects in the 1630s. Merian's investigations were far more thorough than those of her predecessors, as witnessed in the extensive notes that she made. One of the many virtues of the present volume is Etheridge's emphasis on how Merian transformed those notes, selecting and reorganizing them, as she composed the text of her 1679 book. By carefully examining the clues in Merian's published texts (both the 1679 caterpillar book and its 1683 sequel), her notes, her watercolor and gouache studies, and other contemporary sources, Etheridge reconstructs the decades of study that lay behind Merian's first insect book. She shows us that while Merian was also a household manager, mother, teacher, and after the failure of her marriage, entrepreneur, her insect studies occupied a substantial amount of her time.

With this translation and commentary, Michael Ritterson and Kay Etheridge have provided a signal service to historians of science, art historians, cultural historians, and others who wish to reach a deeper understanding of the study of insects in the late seventeenth century, before entomology became a clearly defined scientific specialty. Digitization projects have made several copies of Merian's book available online, but the seventeenth-century German in which Merian wrote and the blackletter typeface in which her book was published will deter many potential readers. Ritterson's translation renders Merian's language into accessible English while retaining the flavor of the original and avoiding anachronistic modern scientific terms. Etheridge's commentary situates each chapter within the context of Merian's larger investigation into nature and later scientific study. Her prefatory chapters offer an excellent introduction to the study of insects before Merian as well as to Merian's life and

---

3 Johnstone was born in Poland to a Scottish immigrant; his name is also written Jan Jonston and Joannes Jonstonus.

work. The concluding chapter traces Merian's posthumous reputation, which was increasingly shaped by publishers' decisions that effaced some of her most noteworthy accomplishments. The volume reproduces a set of exquisitely hand colored counterproof prints of the book, ample evidence that an artist and naturalist like Maria Sibylla Merian could produce accurate scientific illustrations that are also things of beauty.

Preface

... all caterpillars, as long as the adult insects have mated beforehand, emerge from their eggs...[1]

.   .
.

Maria Sibylla Merian (German, 1647–1717) was 32 years old when she published
this statement, written in a period when most people believed that insects
arose by spontaneous generation. Her text on the metamorphosis of silkworms
was reinforced by her carefully etched image of a female moth laying eggs, and
caterpillars hatching and feeding on a mulberry leaf (Plate 1 on p. 148). Raised
in a family of well-known publishers, she was trained from an early age in
painting, etching and engraving. In 1699, she traveled to Surinam to study neotropical insects, and she is today best known for her magnificently illustrated
Metamorphosis Insectorum Surinamensium.[2] Perhaps because of the attention
drawn to the larger and more dramatic book on New World plants and insects,
scholars have tended to overlook her earlier books on the less showy European
organisms. However, these earlier 'caterpillar' (*Raupen*) books, even more than
Metamorphosis, contain a wealth of insightful observations on interactions
among the species portrayed – a cornerstone of the study of ecology.[3]

When I first read the preface to her 1679 book on European moths and butterflies, it took me by surprise. As a biologist exploring a 350-year-old natural
history book, the more I read the more it struck me that this was the work of a
diligent observer, who today would be described as an entomologist working in
behavioral ecology. The perspective that I bring to this book is that of someone
who, like Merian, spent years studying nature in the field and laboratory in
an attempt to satisfy my curiosity about the wonders encountered. Like her, I
have experienced both frustration and pleasure in this pursuit. Perhaps most

---

1   Merian, Maria Sibylla. 1679. *Der Raupen wunderbare Verwandlung und sonderbare Blumen-
    nahrung.* Nuremberg: M. S. Merian: preface.
2   Merian, Maria Sibylla. 1705. *Metamorphosis Insectorum Surinamensium.* Amsterdam: G. Valck.
3   Merian, 1679. *Raupen* and Merian, Maria Sibylla. 1683. *Der Raupen wunderbare Verwandlung
    und sonderbare Blumen-nahrung ... Andrer Theil.* Frankfurt and Leipzig: M. S. Merian. Used
    here for the sake of simplicity, the terms entomology, ecology, biology, and science came into
    use long after Merian.

importantly for this volume, I understand the challenges and the tremendous effort (and, occasionally, luck) required in working with live organisms – even if one has a camera rather than a paintbrush to record visual observations. When I first came across Merian's work many years ago, what puzzled me most was that she had not been acknowledged as the earliest naturalist to combine biologically related plants and animals into one image.[4] Her innovative compositions elevated insects in a new way; they were no longer merely objects to be collected and classified, but actors on the scene, and in a starring role. I am not the first to champion her contributions to natural history by any means, but it seems that some of the earlier voices have been forgotten or ignored, and Merian has received little mention in recent histories of entomology and ecology.[5]

In recent years, scholarship on Merian has shown an uptick, but much of this has emphasized her fascinating life story or her artwork. Some have described her as primarily concerned with the aesthetics of her images. Other scholars have promoted the idea that she was a religious woman who created her books with the first consideration of encouraging piety, and still others have emphasized her work as a part of her livelihood. These motivations were neither insignificant to Merian nor unique to her as a naturalist of the period, but like most of her male counterparts, she was driven primarily by a lifelong curiosity – in her case, about insects. My book was written with the expectation that closer reading of her translated 1679 *Raupen* book and further examination of her information-packed images of insects and their host plants will increase appreciation for the quantity and quality of ecological and behavioral observations that she made.

To understand the significance of Merian's contribution to science, we must acknowledge that prior to Merian's *Raupen* books, animal and plant images were segregated, and these different taxa usually were organized in separate volumes. Natural history studies by her contemporaries often focused on describing and cataloging organisms. However, the systematic ordering of insects was of little or no interest to Merian. After receiving butterfly specimens from English naturalist James Petiver, her response was that she "was not looking for any more specimens, but only at the formation, propagation,

---

4   Etheridge, Kay. 2011. Maria Sibylla Merian and the metamorphosis of natural history. *Endeavour* 35: 16–22.

5   An exception, Frank Egerton includes Merian in his history of ecology with an account of her writings and biography. Egerton, Frank N. 2012. *Roots of Ecology: Antiquity to Haeckel*. Oakland, CA: University of California Press: 194–5. Merian is not mentioned in Worster's foundational work, which like many histories of ecology, begins its focus in the 18th century. Worster, Donald. 1994. *Nature's Economy: A History of Ecological Ideas*. Cambridge: Cambridge University Press.

and metamorphosis of creatures, how one emerges from the other, the nature of their diet ...".[6] Her initial curiosity about their diet sprang from necessity; she wanted to raise insects in order to paint them, and knowledge of their specific food plants was essential to successful husbandry. While some caterpillars will eat almost any herbaceous plant, most are very particular, and they do not survive if deprived of the foods they require. She was the first to write about the significance of this connection between insects and plants, and she depicted these associations in her images of metamorphosis. As exemplified by the cover image for this book, she typically arranged the life cycle of a moth or butterfly around a plant that served as food for the caterpillar. In so doing, she eschewed the traditional format of natural history books that isolated specimens on a page, such that the focus was entirely on their physical structure. Instead, her readers could see and read about how caterpillars were born, fed, grew, and metamorphosed into pupae and then adults. Just as importantly, her text and images were the earliest to show the effects that insects had on the plants that nourished and sheltered them. In addition to the innovative content in her books, she was unusual for her time in that she was responsible for almost every step in the process of production of the 1679 *Raupen* book: she conducted the research, wrote the text, made the images (both drafting the compositions and etching the plates), and was involved in the marketing.

My book was conceived to serve four primary purposes: to provide an English translation of Merian's first scientific work (the 1679 *Raupen* book); to lay out in detail what went into the making of this volume; to analyze and discuss the scientific content of her book; and to address Merian's place in the history of science. The complete title of Merian's 1679 book, like many of its period, reveals much about the work: *The wondrous transformation and particular food plants of caterpillars, wherein, by means of an entirely new invention the origin, food, and changes of caterpillars, worms, butterflies, moths, flies, and other such small creatures, together with their time, location, and characteristics, for the benefit of naturalists, artists, and garden lovers, are carefully investigated, briefly described, painted from life, engraved in copper, and personally published by Maria Sibylla Graff, daughter of the late Matthaeus Merian the Elder*. Parts of this title seemed appropriate as chapter headings for the first part of my book, *The Flowering of Ecology*.

In the first chapter of part 1, *Before the Transformation*, I describe earlier natural history books, which contained the seeds of ecological thought, but portrayed plants and animals separately with little to no discussion of interactions

---

6   Letter from Merian to James Petiver in London, dated 27 April 1705. Folio 70, Sloane 4064. British Library.

among different organisms. An objective of this chapter is to establish the state of 17th century European knowledge on entomology in general, and lepidopterans (moths and butterflies) more specifically.

The second chapter, *A Life Investigated*, recounts Merian's biography and relevant family background, focusing on what is known of her life through the time when she produced the 1679 *Raupen* book. This chapter describes the milieu in which she produced her work, the rich visual and intellectual culture of 17th century Frankfurt and Nuremberg, and the role of women in the artisanal workplace. Her interactions with other naturalists and collectors are addressed, as well as her unique manner of establishing her credentials and authority, while maintaining her position in society as a pious, traditional woman. Lastly, I discuss what I believe were the motivations behind her years of intensive study on insects.

The third chapter, *Described and Painted from Life*, explores the steps of Merian's process in creating her book as far as they are known or can be extrapolated reasonably. I describe her work in the field and in her home 'laboratory,' and her working methods, such as the use of small painted studies to record the metamorphosis of insects. The organization of her caterpillar book is explained and I discuss the composition of the plates, and the ways in which the text and images work together to achieve her goals in presenting her findings on insects and their plant hosts.

In the fourth chapter, *For the Benefit of Naturalists*, I briefly describe the other books that Merian produced over nearly three decades. Central to this chapter is an analysis of the influence of her writings and images on subsequent natural history studies. Ultimately, her ecological compositions created a different way to view nature, and this format was copied so frequently that it became one of the standard ways of composing natural history illustrations. Finally, this chapter addresses possible reasons why the biological information in her work and its significance have been neglected by scholars in spite of her influence and the high regard in which she was held by her near contemporaries.

The second part of this volume, *Maria Sibylla Merian's Caterpillar Book*, presents an annotated English translation of the complete 1679 *Raupen* book text integrated with my commentary. As described in the translation notes, Michael Ritterson and I have attempted to retain the flavor of Merian's style and language, and we have adhered to English vocabulary that reflects what was known about insects and plants in her time. As in the original book, the text is accompanied by her images of the metamorphic stages of insects with the plants on which the caterpillars usually feed. My commentary provides an overview of the natural history content of both the text and images in her book, and I discuss the novel information in her accounts of the insects she raised and observed. Special attention is given to her study of factors central to ecological science, including her descriptions of environmental effects on insect development and abundance and observations on insect food choice and feeding behavior.

Merian published her books on insects at a critical point in European natural history studies, and the *Raupen* books and *Metamorphosis* changed how nature was described and portrayed. Almost two centuries after Merian's death, the study of insect–plant interactions became a focal point in the growing fields of ecology and evolution, and thousands of papers and books have been dedicated to these critical relationships. Plants comprise more than half of the biomass on Earth, and insects are the most prevalent animals on the planet in terms of total biomass. Thus, plant–insect interactions are a key to understanding terrestrial ecology. Wherever there are plants, there are insects set on using them for their own devices, but conversely, many plants employ insects for their benefit as well, primarily as pollinators. Lepidopterans in particular show two faces to the world: the larval stages (caterpillars) are perceived primarily as destructive and by most, ugly or repellent, whereas adult butterflies and moths are often thought of as beautiful, flying through the landscape adding color and motion. Caterpillars are among the most ravenous and destructive herbivores in many terrestrial ecosystems, capable of consuming three times their body weight in a day. As such they are the bane of farmers and gardeners the world over, yet they are essential elements of many a food chain, eaten by countless other animals. The adult stages frequently serve plants as pollinators as well as food for other animals.

Merian as a naturalist and an artist had a life-long love affair with all stages of the lepidopteran life cycle. She pursued her passion for more than five decades, recording the food plants, behavior and ecology of roughly 300 species of insects and publishing four books on her scientific work. Much of what she described withstands the scrutiny of modern biology, and the number of original observations in her work are astounding in an era when most biologists focus on one or two organisms in their career. Her images of insects on plants create small dramas, drawing attention to these critical links in the ecosystem in a way that was revolutionary for her time. Merian's years of work generated a fundamental shift in the trajectory of natural history studies, and it is well past time that this was recognized.

Acknowledgments

When I was five or six years old, my father gave me an illustrated book on the natural history of animals and plants. I remember looking through it with him, fascinated by the images and curious about the lives of the animals pictured. As my interest in nature grew, my mother gave me jars for pet insects and tolerated the tadpoles, snakes and salamanders I brought home. It seems that the path towards making this book was set at an early age. Becoming a biologist was a gradual education with the usual formal elements. Learning how to investigate the history of science and natural history art has come about by a more circuitous route, assisted along the way by countless people. I regret that I cannot recognize all of them by name here.

I must begin by thanking my friend and translator, Michael Ritterson. Translation of the *Raupen* text was a long, iterative process as we learned more about how Maria Sibylla Merian expressed herself, the idiosyncrasies of 17th century German, and the biological terms in use (or not) in her day. Michael's patience and attention to detail are extraordinary, but he contributed much more than files of translated text. He looked over the primary and secondary literature in German that I came across, helped determine what was of interest to this project, and translated these sources as needed. In the process, he independently found other sources, and in one case, contacted the author of an article, Margot Lölhöffel, for more information. She very kindly responded with added information and more articles. Thanks to Michael, Margot became my correspondent, friend, and ultimately, my primary source for learning about the history of Nuremberg and Merian's life and work in that city. She and Dieter Lölhöffel hosted me at their home, showed me Merian's Nuremberg, and together we made discoveries in the rich library and museum collections in that city as well as in nearby Bamberg and Erlangen. A highlight of my time with Margot and Dieter was a visit to 'Merian's garden' in grounds of the Nuremberg castle. Thanks to their tireless efforts to bring Merian's history to the eye of the modern public, this restored garden has been dedicated to honoring her work there.

Florence Pieters, former curator of the Artis Library, has been an amazing font of knowledge, as well as my friend, supporter, and sounding board for more than a decade. Florence and I have spent many enjoyable hours vigorously discussing all things Merian. Her book-filled home in Amsterdam, which we have dubbed the 'Maria Sibylla Merian Study Center,' has been my base for weeks of work at the Artis Library, the Rijksmuseum and places slightly further afield such as Teylers Museum in Haarlem.

At the Artis Library, I came to know Hans Mulder, current curator of this rich collection of natural history literature and images. Hans became another irreplaceable resource and a pillar of unwavering support during my ever-expanding Merian project; he also facilitated the acquisition of many of the images for this book. Florence and Hans have contributed endlessly to what I have learned about Merian and her time in Amsterdam, but also about the history of books. Both have translated Dutch text for me, some excerpts of which are in the present volume. Hans also introduced me to Ad Stijnman, whose meticulous tutelage on the process of etching and engraving clarified a good many pieces of the Merian puzzle. Through Florence and Hans, I came to know other experts in natural history studies and the social/scholarly networks of my period of interest. I especially want to thank Diny Winthagen, Bert van de Roemer, and Marieke van Delft, all of whom have lent a hand to the project, both through their knowledge and their friendship.

Over time, my developing 'Merian network' led me to meet and correspond with Kurt Wettengl and Katharina Schmidt-Loske, whose scholarship and valuable insights on Merian further informed my research. Through Florence and Diny, I had the good fortune to meet Joos van de Plas, an artist who has investigated Merian in a number of extraordinary ways in her own work. Joos has raised many species of butterflies and moths through metamorphosis as part of her artwork. As I visited her studio and garden, she led me to understand the amount of work and care involved in lepidopteran husbandry, as well as providing an utterly fresh perspective from which to view Merian's work.

Two other excellent teachers on my way to learning about natural history books and their genesis have been Leslie Overstreet and Tony Willis. Leslie assisted me extensively as I worked in the Cullman Library of the Smithsonian, and she has answered endless questions over the years both in person and by email. She and Janice Neri, whose scholarship also inspired me, were essential in helping me obtain much needed funding by writing letters of recommendation to some of the organizations listed below. Tony Willis guided me as I explored the unique collection of books, manuscripts, and artwork at the Oak Spring Garden Library (e.g. Jacob Marrel's tulip portraits), and he has graciously provided images for my book. Charlotte Tancin of the Hunt Institute for Botanical Documentation has answered my questions on a number of occasions, helping me to understand the use of printers' ornaments as just one example. This book was made possible by many more institutions and their staff than can be named here, and I am infinitely appreciative. I was repeatedly amazed by the helpful and cooperative spirit of people who steered me to resources in their collection, answered my many questions, and facilitated scans and permissions for the dozens of illustrations that appear in this book.

Once the book manuscript was in draft form, another team of talented friends and colleagues came to my assistance. The book has benefited tremendously from Esther van Gelder's careful reading of more than one draft. She provided a number of thoughtful suggestions that greatly improved the structure of the book. Esther seemed to discover every single place in the manuscript where I had struggled with organizing large amounts of information, and she proposed an elegant solution for each of these tangles. Dominik Hünniger read early versions of the German translations and made constructive suggestions, as did Florence Pieters. I am indebted to Brian Ogilvie for his detailed feedback on the manuscript and for providing the foreword; his extensive writings also steered much of my thinking about material in the first chapter. Andreas Weber was very helpful as my series editor in bringing the final stages of the manuscript to publication, and at Brill, Rosanna Woensdregt provided ever-patient assistance with publication details. Throughout the manuscript phase of the project, Henrietta McBurney Ryan provided both wisdom and moral support at a number of times when I felt like this book would never be finished.

I am grateful for funding support from Gettysburg College, the Renaissance Society of America (Paul Oskar Kristeller Grant), and the American Philosophical Society (Franklin Research Grant). A fellowship from the National Endowment of the Humanities supported an extended sabbatical at a critical point in the project.

Last but certainly not least, I am grateful for the encouragement and patience of family and friends, especially my husband, Don Walz.

Illustrations

Figures

Plates from Merian's 1679 *Raupen* Book

PART 1

The Flowering of Ecology

∴

CHAPTER 1

Before the Transformation

Introduction

Much has been written about the interwoven evolution of art and natural history in the Renaissance, but a brief overview with a focus on insects will establish what laid the foundation for Maria Sibylla Merian's work, and how her first book on insects departed from those that preceded it. As we shall see, the 17th century was a time of booming interest in insects among European scholars and amateur collectors as well as in the art world. Here I will examine some of the driving forces behind this phenomenon. Plant natural history and images[1] will be considered, but to a lesser extent, because these organisms played a supporting role in Merian's research. The emphasis is on published natural histories, both because that is what she produced, and because printed works were used to disseminate information more widely than was possible with individual copies of an image or text. In order to examine the place occupied by Merian's work on insects in the progression of natural history studies, it will be essential to address the interplay of art and science, the influence of religious beliefs, and the practical challenges of publishing heavily illustrated books.

The metamorphoses of insects, in particular moths and butterflies, are central to both the images and the text of Merian's books. Others before her investigated metamorphosis to varying degrees, but she was the first to tie the process to the required foods of the larval insects. To set the stage for what follows, it is helpful to compare a series of images from books that preceded Merian to one of her representative plates. Prior to her 1679 *Raupen* book, plants and insects were set apart in unrelated books (Figures 1 and 2).[2] Furthermore, before the

---

1 My choice of the word image rather than art or illustration is intentionally neutral. Merian's work is an apt example: she has been considered an excellent artist both in her time and ours, but her images, while composed with an artist's eye for aesthetics, were structured to illustrate her observations on insects. For more background on natural history images in the period considered here, see Kusukawa, Sachiko. 2011. Patron's review. The role of images in the development of Renaissance natural history. *Archives of Natural History*, 38(2): 189–213 and Dickenson, Victoria. 1998. *Drawn from Life: Science and Art in the Portrayal of the New World*. Toronto: University of Toronto Press: 6–10.

2 See for example Mattioli, Pietro A. 1568. *I Discorsi di M. nelli sei libri di Pedacio Discodoride Anazarbeo della materia medicinale*. Venezia: Vincenzo Valgrisi; Moffet, Thomas, Edward

© KONINKLIJKE BRILL NV, LEIDEN, 2021 | DOI:10.1163/9789004284807_002

FIGURE 1   Woodcut showing a nettle plant. Mattioli, 1568. *I Discorsi*,
            p. 1126. This edition was printed on blue paper.

publication of Johannes Goedaert's *Metamorphosis naturalis* in which he combined the metamorphic stages into one plate (Figure 3),[3] these life stages were
usually treated separately, with adults and larvae appearing in different parts
of a book (Figure 2). Merian moved research on metamorphosis a large step
forward by combining the complete metamorphic cycle of insects with ecologically related plants (Figure 4). Her text worked with the images to emphasize

        Wotton, Conrad Gessner, Thomas Penny, and Théodore Turquet de Mayerne. 1634. *Insectorum
        sive Minimorum Animalium Theatrum*. London: Thomas Cotes.
3   Goedaert, Johannes and Johannes de Mey. 1660–1669. *Metamorphosis Naturalis*. Middelburgh:
        J. Fierens.

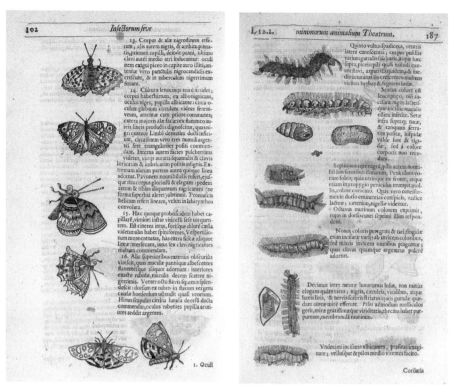

FIGURE 2   Woodcut images of adult moths and butterflies along with descriptive text (left).
Moffet et al., 1634. *Insectorum Theatrum*, p. 102. Larval forms are shown on p. 187 of the
same volume (right).

this critical relationship. Moreover, inclusion of a plant provides a framework
for illustrating insect behaviors and movement.

Art and Science Intertwined

Both Merian and Goedaert were naturalists who originally trained as artists,
and many natural philosophers were skilled in using the techniques of art to
illustrate their work. The perceived gulf between artist and scientist as these
terms are narrowly defined today would have been puzzling to scholars in
16th and early 17th century Europe. The Latin roots of the words are helpful
here; *ars* (art) embraced things created by man rather than by nature, and
scientia (science) referred to accumulated facts and knowledge. Art and science (including natural history as well as physics and chemistry) are still ways

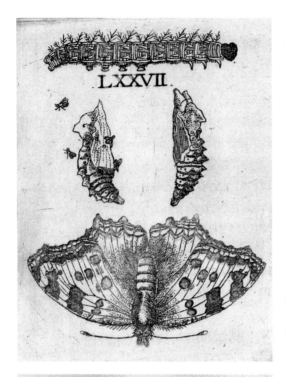

FIGURE 3
Three stages of metamorphosis of
Aglais urticae (small tortoiseshell
butterfly). Goedaert, 1660.
Metamorphosis Naturalis, Plate 77

FIGURE 4
The metamorphic cycle of *Aglais
urticae* (small tortoiseshell
butterfly) on a nettle plant, the
required food of the caterpillar
of this species. Merian, 1679.
Raupen, Plate 44

of knowing the world; however, from the Middle Ages through the Renaissance and into the Early Modern Period these means of understanding were inextricably linked. It has been convincingly argued that Renaissance artists and artisans were key players in the shift from the received wisdom of classical scholars to active engagement with nature, and that the movement toward modern science was tied to the history of naturalism in art.[4] Of course, naturalistic images preceded the Renaissance by millennia, appearing for example in Greek and Roman art and decoration. Not only plants, birds and mammals, but also 'lowly' insects were depicted in luxuriously decorated margins of medieval manuscripts (Figure 5); these and other Renaissance images heralded the 'rebirth of naturalism' rather than its invention.[5] Continuing well into the Early Modern period, plants and animals continued to serve as decoration and symbols, but their rendering often provided surprisingly accurate depictions.[6] In medieval herbals, images of plants took on an additional role, providing information for identification, essential in compounding medicines and other potions. Numerous illustrated herbals existed long before the advent of printing. Eventually, naturalists began to use these information-laden images to name and order plants, and a multitude of herbals and florilegia were published from the 16th century onward. The study of plants was a serious endeavor long before naturalists turned their eye to insects, which rarely received specific names before the 18th century.

---

4   Smith, Pamela H. 2004. *The Body of the Artisan: Art and Experience in the Scientific Revolution.* Chicago: University of Chicago Press: 19. Smith includes the art of Dürer and life-casts crafted by Palissy as embodying her well-founded premise that "knowledge of nature is gained through direct observation of particular objects and that nature is known through the hands and the senses rather than through texts and the mind." This has been similarly argued in Ackerman, James S. 1985. The involvement of artists in Renaissance science. In *Science and the Arts in the Renaissance.* John W Shirley and F. David Hoeniger, Eds. Washington, DC: Folger Shakespeare Library: 126 and in Topper, David R. 1996. *Towards an Epistemology of Scientific Illustration.* Toronto: University of Toronto Press: 225–27.

5   Blunt, Wilfrid and William Thomas Stearn. 1950. *The Art of Botanical Illustration: An Illustrated History.* North Chelmsford, MA: Courier Corporation: 15.

6   Art historians tend to prefer to describe such images as naturalistic rather than accurate, but here I use the latter term to denote an image that contains visual details rendered in a way that makes an organism identifiable to a biologist. An image of a butterfly can be naturalistic without containing the correctly rendered details to make it recognizable as a specific type of butterfly (by taxonomic family if not genus or species).

FIGURE 5   Attributed to Simon Bening (Flemish, circa 1483–1561). *Horae Beatae Mariae*
Virginis ad usum Romanum. Illuminated manuscript on vellum. Folio 103v and
folio 104r

Insects in Drawings and Manuscripts

Like plants, insects are ubiquitous in all environments inhabited by humans,
and of all animal groups, these small creatures have the greatest effects on society, playing a role in health, agriculture and economies. But study of this largest group of animals, almost 40% of all animal species, was not a top priority
for the earliest natural philosophers. Only snakes, which like worms and other
crawling creatures were once included in the group called *insecta*, were lower
on the great chain of being, a hierarchy attributed to God.[7] Popular opinion
of insects was not elevated by stories that blamed them for Biblical plagues
and attributed their formation to spontaneous generation from mud and dung.
Curiosity about insects and their metamorphosis was evident in the writings
of the classic authors, but published images of insects did not appear until well

7   Today, the scientific use of the word insect is confined to arthropods in the class Insecta, but
in the period of interest here that was not the case. For the sake of simplicity, I use insect in
the modern sense throughout this book.

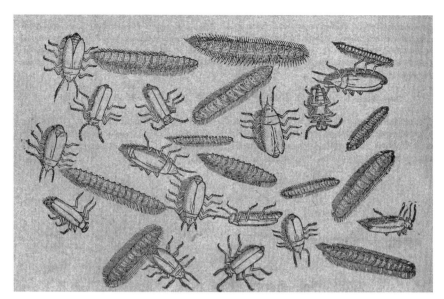

FIGURE 6   Woodcut image of the pine processionary moth and unknown beetles. Mattioli,
1568. *I Discorsi*, p. 382

after those of plants. One of the earliest printed images of caterpillars appears
in 1568 in Pietro Mattioli's plant-oriented medical treatise (Figure 6), which
also features cicadas and bedbugs.[8] However, publication dates of printed
works do not tell us when the first studies were made on a group of organisms.

A number of collections of naturalistic insect drawings and paintings are
known from late 16th and early 17th century in Europe, and undoubtedly,
many more have been lost to history. Some of these images eventually were
transferred to woodblocks or copperplates and printed, but many more were
made for private study and were not published. Still, they are important in-
dicators of dedicated interest in a group of relatively overlooked organisms.
Adriaen Coenen (or Coenensz, 1514–1587), the son of a Dutch fisherman,
collected naturalia, compiled information, and made images predominately
related to the sea, the coast and marine life.[9] Coenen may not have been aca-
demically trained, but he copied other authors, indicating that he was aware
of the work of scholars. Most interestingly, he made meticulous first-hand

8   Mattioli, 1568. *I Discorsi*: 382. The caterpillar of the pine processionary moth (*Thaumetopoea
    pityocampa*) was known as *Brucha di pini*, and is famous for its urticating hairs that can cause
    severe pain.
9   Coenensz, Adriaen. 1577–1580. *Vis booc*. The manuscript is in the Jacob Visser collection,
    National Library of the Netherlands. The images are watercolor on paper and are inter-
    spersed with writing.

FIGURE 7   Adriaen Coenensz, 1577–1580. *Vis booc*. Left-hand side folio 309v and right-hand side folio 314r

FIGURE 8   Adriaen Coenensz, 1577–1580. *Vis booc*. The left image shows details from folio 309v above, and the right image of metamorphosis is a detail from folio 314r.

observations of a number of natural phenomena, including the stages of metamorphosis of both frogs and insects. In one image he shows a plant covered by caterpillars that appear to have eaten all of the leaves (Figure 7). In a nearby sheet he shows the stages of metamorphosis of a caterpillar in a series of 'time-lapse' vignettes (Figure 8).[10]

10   Egmond, Florike. 2017. *Eye for Detail: Images of Plants and Animals in Art and Science, 1500–1630*. London: Reaktion Books Limited: 200–8. Egmond provides an excellent overview of techniques used to represent nature in the 16th century.

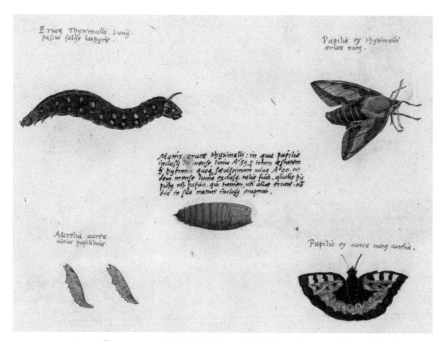

FIGURE 9   Caterpillar, pupa and adult of *Hyles gallii* (bedstraw hawkmoth) and handwritten
text in Latin. Felix Platter album, watercolor on paper, Folio 113

Unlike Coenen, Felix Platter (Swiss, 1536–1614) was formally educated as a
physician. For decades, he collected plant and animal images, which were
cut and pasted into albums and annotated. Among the pages of insects are
images of caterpillars on some pages and adult moths and butterflies on
other sheets, but there are indications that Platter studied metamorphosis.
One page includes the caterpillar, pupa and adult of the readily identifiable
bedstraw hawkmoth (*Hyles gallii*). The text above the pupa dates the observation of this metamorphosis to 1589–1590 (Figure 9). The dates fit with the
fact that the species overwinters as a pupa and the moth emerges in the
next year. Platter published medical articles but nothing from these albums
of naturalia, and the images remained relatively unknown until recently.[11]
Decades later, Alexander Marshal (English, c. 1620–1682), depicted dozens
of insects in careful watercolors and made brief notes on some of these. He
shows metamorphic stages of lepidopterans in several images, but his studies
were revealed to the world only in 1980 (Figure 10). Like Marshal, John White
(English, 1540–1593) never published his images of North American insects,

---

11   Ibid.: 212–17.

FIGURE 10
Watercolor on paper of
Aglais io (peacock butterfly)
adult, pupa and caterpillar.
Alexander Marshal. Folio 4r.
Merian's image of this
species is on p. 254 of this
book.

but some of his material was published by other naturalists.[12] The aforementioned individuals engaged in the process of "understanding the natural world through image making," a hallmark of insect study in this era.[13]

The nature images made by observers of nature such as Platter, Marshal and White were created with a different purpose from the lifelike insect paintings made by the Flemish miniaturist Joris Hoefnagel (1542–1601). Hoefnagel did not study his subjects with an eye to understanding their reproduction or other natural functions, but in order to be able to paint them as accurately as possible. Beginning in the 1570s and continuing for the rest of his life, he created hundreds of lifelike miniatures on vellum of animals ranging from whales

12   Leith-Ross, Prudence and Henrietta McBurney. 2000. *The Florilegium of Alexander Marshal: In the Collection of Her Majesty the Queen at Windsor Castle*. London: Royal Collection: 15–19. Harkness, Deborah E. 2009. Elizabethan London's naturalists and the work of John White. In *European Visions – American Voices*. Edited by Kim Sloan. London: British Museum: 44. White shared insect images with Thomas Penny and Thomas Moffet, whose work led to the creation of the 1634 *Insectorum Theatrum*, which is discussed below.

13   Neri, Janice. 2011. *The Insect and the Image: Visualizing Nature in Early Modern Europe, 1500–1700*. Minneapolis: University of Minnesota Press: xi–xiii. Neri proposed that by making natural subjects such as insects into visual objects available for study, the artists became "gatekeepers to a strange and fascinating world."

FIGURE 11   *Papilio machaon* (old world swallowtail). Joris Hoefnagel, 1575/1580. Watercolor
and gouache on vellum. Folio 7 in *Animalia Rationalia et Insecta* (*Ignis*)

to arthropods. Many of the larger animals were copied from other sources such
as Conrad Gessner's animal encyclopedias,[14] and were greatly reduced in size
to fit onto a small sheet of vellum. However, his insects and spiders are usually shown as life-sized, and their accurate portrayal indicates that most were
painted from direct observation of living or preserved animals (Figure 11). Over
a number of years, Hoefnagel depicted a much greater variety of insects and
arachnids than any Western artist before him.[15]

    However precise his observational skills, we have no evidence that Joris
Hoefnagel made natural history notes on the animals that he painted; instead,
his goal was to attribute meaning to the animals depicted. The beautifully
rendered images are adorned with text alluding to the symbolic nature of the

14   Gessner, Conrad. 1551–1558. *Historia Animalium*. Five volumes. Zurich: C. Froschauer.
15   These four Hoefnagel manuscripts, titled in English the *Four Elements*, are in the col-
      lection of the National Gallery of Art, Washington, D.C. Volume 1 consists of *Animalia
      Rationalia et Insecta* (*Ignis*). For more on Joris Hoefnagel's miniatures and the transla-
      tion of some of these into print, see Vignau-Wilberg, Thea and Hoefnagel, Jacob. 1994.
      *Archetypa Studiaque Patris Georgii Hoefnagelii, 1592: Natur, Dichtung und Wissenschaft
      in der Kunst um 1600: Nature, Poetry and Science in Art around 1600*. Berlin: Staatliche
      Graphische Sammlung.

creatures portrayed; this Latin text is taken from a variety of sources, including the Bible, Erasmus[16] and various classic texts such as Ovid's *Metamorphosis*. The miniatures were begun in the early 1570s at Hoefnagel's own initiative and with no patron. This image collection must have been important to him as a personal means of reflection and expression, because he kept it with him throughout extensive travels and for his entire life, constantly adding images and epigrams to the manuscript.[17] Ultimately, Hoefnagel's private collection of painted insect images had influence beyond the more scientific observations enumerated above, for the simple reason that many were printed. The wide dissemination of Joris Hoefnagel's insect images was implemented by the publication of *Archetypa Studiaque Patris Georgii Hoefnagelii* in 1592, one of the first set of plates dedicated to small animals integrated with parts of plants.[18] Joris credited his son Jacob Hoefnagel (Flemish 1573–1632) with engraving the plates, but variation in the skill and delicacy with which they were executed make it likely that more than one engraver was involved in the project. The animals depicted usually were insects and other arthropods (including a crab), but mollusks, reptiles, amphibians, a mouse and even a bird embryo feature in this eclectic mix of images. On most plates, there are two or more flowers, some fruits and roots, and a few small floral arrangements are included (Figure 12). As in the still life paintings that these plates prefigure, no biological relationship among the organisms in an image can be detected. The publishing history of various editions of the *Archetypa* is complex, but the fact that it was reprinted several times increased the availability of the plates. The influence of this volume and subsequent editions and variations led to wide use of its images as models for insect encyclopedias such as that of Jan Jonston[19] (Figure 13) and for 'artist-naturalists' such as Maria Sibylla Merian.[20]

---

16 Erasmus, Desiderius. 1508. *Erasmi Roterodami Adagiorum Chiliades Tres*. Venice: Aldus Manutius. Thousands of proverbs were compiled and annotated by Erasmus and published in numerous editions.

17 Bass, Marisa A. 2017. Mimetic obscurity in Joris Hoefnagel's Four Elements. In *Emblems and the Natural World*. Edited by Karl A. E. Enenkel and Pamela J. Smith. Leiden: Brill: 526–38. For more on the context of Hoefnagel's work and his motivations, see Bass, Marisa Anne. 2019. *Insect Artifice: Nature and Art in the Dutch Revolt*. Princeton: Princeton University Press.

18 Hoefnagel, Joris and Jacob Hoefnagel. 1592. *Archetypa Studiaque Patris Georgii Hoefnagelii*. Frankfurt: Jacob Hoefnagel.

19 Jonston, Jan, Matthaeus Merian, and Caspar Merian. 1653. *Historiae naturalis de insectis, Libri III*. Frankfurt-am-Main: Merian.

20 Vignau-Wilberg and Hoefnagel, *Archetypa*: 13–20 and 85–6. Jacob Hoefnagel was just 19 at the time of publication of the plates. Vignau-Wilberg speculated about the possible involvement of the Frankfurt-based Theodor de Bry publishing firm, for whom Maria

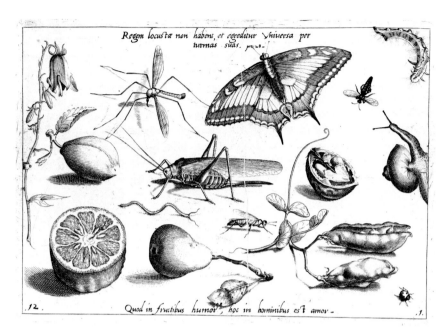

FIGURE 12   Joris Hoefnagel, 1592. *Archetypa*, Part 1, Plate 12. The *Papilio machaon* (old world
                   swallowtail) was painted more than a decade earlier by Joris Hoefnagel. The twisting
                   caterpillars in this plate were echoed in a number of 17th century still life paintings.

FIGURE 13
Plate 6 in Jonston and Merian, 1653. *Naturalis
de Insectis Libri III*. The *Papilio machaon* (upper
left) was copied from Hoefnagel's *Archetypa*
(Figure 12 above).

Still Life Painting

Naturalists and artists such as Joris Hoefnagel often isolated organisms on a page in order to focus on their individual traits, a way of presenting nature termed 'specimen logic' by Janice Neri.[21] In contrast, artists of the 17th century often combined a profusion of insects and plants, particularly flowers. The genre of floral still life painting had its roots in earlier flower paintings such as the border decorations of medieval and Renaissance prayer books (see Figure 5).[22] Merian's compositions combining plants and insects might appear at first glance to spring from the tradition of floral still life painting, but the relationship is superficial. Floral still life paintings are laden with symbolism while being largely devoid of naturalistic relationships. The plants depicted may have had their origins on separate continents and often flowered at different times of the year. The frequently depicted insects and other small animals appearing in these paintings had no relationship to each other or to the flowers other than the fact that they might bear similar meaning for the viewer. In some cases, still life compositions were designed to showcase naturalia and ornamental items collected by the person who commissioned a painting. The neat ordering of European nature experienced a tremendous upheaval by the beginning of the 16th century. Exploration and the ensuing globalization of trade resulted in the flow of exotic plants and animals, living and dead, into the hands of the curious. Explorers and traders brought home marvels that made their way into gardens, menageries and collections of those who could afford sunflowers from North America, a nautilus shell or a rare butterfly specimen. These marvels often appeared in paintings commissioned by the collectors.

Of particular interest here is the predominance of insects in still life paintings. Like flowers, insects are short-lived organisms that represented the transient nature of life. Insects also served as symbols of death (a fly), the eternal life of the soul (a damselfly or dragonfly), or the resurrection of Christ (a butterfly). The presence of these small creatures in paintings peaked in the 17th century, when as many as four out of five such artworks contained insects.[23]

Sibylla Merian's father later worked as an engraver. The plates were sold to Nuremberg publisher Paul Fürst (1608–1666), who also published them around 1665. Vignau-Wilberg recounts the complex publishing history of subsequent editions and copies, which appeared through the 18th century.

21   Neri, *Insect and the Image*: xxii.

22   Wheelock, Arthur K. 1999. *From Botany to Bouquets: Flowers in Northern Art*. Washington: National Gallery of Art: 16.

23   Dicke, Marcel. 2000. Insects in Western art. *American Entomologist*, 46(4): 228–36. The 17th century peak was determined from a survey of 1,942 pieces of art from the 15th to the

Hoefnagel's *Archetypa* likely spurred the use of detailed images of insects in northern European art; this 1592 book and its later renditions were used extensively as model books by artists like Georg Flegel (German 1566–1638), a master of naturalistic plant and insect portrayals, and his student, Jacob Marrel (German, 1613–1681), Maria Sibylla Merian's step-father and teacher (Figure 14). At the same time that insects were increasingly painted into works of art, they were being examined more and more by naturalists, and in a variety of ways. Insects were small, easily preserved and highly portable collectibles. By the 17th century, they were increasingly popular items in the collections of scholars and amateurs alike.[24] To understand fully the growing interest in insects as objects worth of collection and study, we must further investigate their role as symbols and emblems.

Finding God in Nature

In considering how plants and animals have been perceived in Western culture, one cannot separate the development of natural history studies from the evolution of religious beliefs. Just two aspects of this complex topic will be addressed here. First is the practice alluded to previously: the use of organisms to represent ideas and allegories; the second is the development of natural theology, the belief that the study of nature honors God. The use of a lion to represent courage and a snake to portray evil are familiar themes even today, so emblematic animals are not unknown to us. A limited cast of insects served as symbols and played roles in moral allegories in classic writings, the Bible and bestiaries (bees are industrious and locusts deliver retribution). The *Four Elements* albums by Joris and later Jacob Hoefnagel's *Archetypa* included epigrams with each image, and these alluded to some characteristic of one of the plants and animals portrayed. As in still life paintings and the even earlier illuminated manuscripts, flowers could stand in for the ephemeral nature of beauty, riches or life, and the slow progress of a snail represented a caution

---

20th century. Dicke estimates that 82% of the still life paintings in this period contained insects, and of these, 77% were moths, butterflies or their caterpillars.

24 The techniques used for collecting and preserving insects were increasingly documented by naturalists in the 17th and 18th centuries. Hünniger, Dominik. 2018. Nets, labels and boards: materiality and natural history practices in continental European manuals on insect collecting 1688–1776. In *Naturalists in the Field: Collecting, Recording and Preserving the Natural World from the Fifteenth to the Twenty-First Century*. Edited by Arthur MacGregor. Emergence of Natural History, Volume 2. Leiden: Brill: 686–705.

FIGURE 14   Jacob Marrel, 1634. Still life with a vase of flowers and dead frog. Oil on wood
             panel. Interestingly, Marrel included two caterpillars but no adult moths or
             butterflies in this composition.

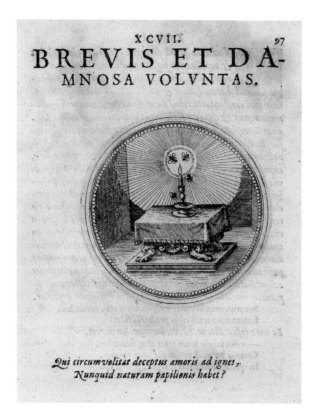

X C V I I. 97
B R E V I S E T D A-
MN OSA VOLVNTAS,

Qui circumvolitat deceptus amoris ad ignes,
Nunquid naturam papilionis habet?

FIGURE 15
Moths are shown
being drawn to a
flame. Camerarius,
1596/1604. *Symbolorum &*
Emblematum, Volume 3,
Plate XCVII

against rash haste.[25] The tradition of insects and other organisms as emblems was on rich display in four illustrated volumes by Joachim Camerarius the Elder (German, 1534–1598), the third of which contained birds and insects (Figure 15). Most of his emblematic associations follow those of Pliny.[26] However, the appearance of organisms as symbols from nature was not limited to specialized emblem books and paintings, but remained embedded in natural history treatises well into the 17th century. In Gessner's animal encyclopedia for example, emblematic natural history was integrated with the writings of classic authors and fresh observations. Of the sixteen pages on the fox in his

---

25 Vignau-Wilberg and Hoefnagel, *Archetypa*, 55–7. The sources of the epigrams include quotations from the Bible, Ovid and other classical sources, Erasmus and possibly original poems and riddles from Joris Hoefnagel.

26 Camerarius, Joachim. 1596/1604. *Symbolorum & Emblematum ... centuria vna* [-quarta], Nuremberg: [part 1] for Johann Hofmann and Hubert Camoxius, 1590 [1593]: [part 2] P. Kaufmann, 1595: [part 3] ibid. 1596 [1597]: [part 4], 1604.

encyclopedia, six were devoted to biblical and other literary references, as well as the linguistics, myths and fables associated with the animal.[27]

One could learn morality from contemplation of the allegories that flowed from emblematic nature, but these lessons relied upon received wisdom and this approach to natural history, although important to the humanists working in the field, was unlikely to lead to new discoveries. The growth of knowledge based on direct observation of nature, as conducted by Maria Sibylla Merian and her near contemporaries, also had a spiritual element however; this endeavor was the search for the wisdom inherent in God's creations. Natural theology is the study of nature as a way to know God. Like emblematic nature, it had its seeds in the writings of the ancients, and for the Renaissance scholar influenced by the classics, no conflict existed between the pursuit of 'science' and the state of one's soul. To understand the range of nature from a tiny flea to the order of the cosmos was to understand God, and provided a way of melding human curiosity with the major belief systems of the day.[28] Conrad Gessner (Swiss, 1516–1565) included the historical emblematic meanings for creatures included in his animal encyclopedias, but also added a theological dimension. In his writings he made it clear that he contemplated nature in order to understand its creator. Gessner considered the description of nature, as in his *Historiae animalium* (1551), as a way to praise God and to interpret parts of the Bible, in particular *Genesis*.[29] Early naturalists such as Gessner were "reading endlessly from the book of nature, in an attitude of humility dictated by the constant appearance of new signs of limitless divine power."[30] It has been proposed that by the 17th century, the pursuit of God through nature

---

27 Ashworth, William B. 1996. Emblematic natural history of the Renaissance. In *Cultures of Natural History*. Edited by Nicholas Jardine, James A. Secord, and Emma C. Spary. Cambridge: Cambridge University Press: 18–21 and 34–5.

28 For the influence of natural theology on natural history studies see Jorink, Eric. 2010. *Reading the Book of Nature in the Dutch Golden Age, 1575–1715*. Leiden: Brill, and Ogilvie, Brian. 2005. Natural history, ethics, and physico-theology. In *Historia: Empiricism and Erudition in Early Modern Europe*. Edited by Gianna Pomata and Nancy G. Siraisi. Cambridge, MA: MIT Press: 75–103. Natural theology has a long history predating Christianity. In the 17th and 18th centuries a similar but more specific concept termed physico-theology was in use, often by Protestant scholars. See Blair, Ann and Kaspar von Greyerz (Eds.) 2020. *Physico-theology: Religion and Science in Europe, 1650–1750*. Baltimore: Johns Hopkins University Press.

29 Enenkel, Karl A. E. 2014. The species and beyond: classification and the place of hybrids in Early Modern zoology. *Zoology in Early Modern Culture: Intersections of Science, Theology, Philology, and Political and Religious Education*. Leiden: Brill: 85.

30 Glardon, Philippe. 2007. The relationship between text and illustration in mid-sixteenth-century natural history treatises. In *A Cultural History of Animals in the Renaissance*. Edited by Bruce Boehrer. Oxford: Berg: 140.

was virtually a mandate for learned Protestants in northern Europe.[31] Many
naturalists were theologians or clerics, including the botanist Otto Brunfels,
(German, 1488–1534) and John Ray (English, 1627–1705). Some lesser known
theologians stayed in the background but were highly influential; one example
is that of the Middelburg preacher Johannes de Mey (1617–1678), who aided
Johannes Goedaert in numerous ways in the publication and promotion of the
first book dedicated to elucidating the metamorphosis of insects.[32] The vibrant
interest in natural theology had not waned by the middle of the 17th century,
and insects were increasingly emerging as objects of interest for collection and
investigation. Pliny the Elder's 1st century CE writings on insects were cited for
almost a millennium and a half, and his statement that Nature can be seen "nowhere more than in her smallest creations"[33] was a long-standing argument in
support of their study.

Natural History of Insects in Printed Works

The natural histories of the ancients such as Aristotle, Pliny and Dioscorides
were passed on for centuries in manuscript form, and their observations were
central to the earliest printed works on plants and animals.[34] Aristotle used
the term *entomon* ('notched') for insects and developed a simple classification
scheme for them. Roughly 500 year later in his *Historia Naturalis* Book XI, Pliny
added to Aristotle's work, describing more types of insects.[35] With the advent
of printed books, classical authors were quoted extensively by naturalists like
Gessner and Ulisses Aldrovandi (Italian, 1522–1605), who assembled information on nature from a wide array of sources. Their early animal encyclopedias
included etymology, emblematics, and efforts at classification, as well as some

---

31   Trepp, Anne-Charlott. 2005. Nature as religious practice in seventeenth-century
      Germany. In *Religious Values and the Rise of Science in Europe*. Edited by John Brooke and
      Ekmeleddin İhsanoğlu. Istanbul: Research Center for Islamic History, Art and Culture:
      81–110. Trepp explored religious attitudes and practices as well as natural history study in
      the context of 17th c German Protestantism. She argues that the pluralism that developed
      in 17th c Lutheran theology was linked to an increase in interest in natural history; this
      link was generated by the development of a more "practice-oriented form of piety."
32   Goedaert, *Metamorphosis Naturalis*. Three volumes were issued in Dutch and also in
      Latin. For the role of De Mey in Goedaert's work see Jorink, *Book of Nature*: 205–7.
33   Rackham, Harris. 1940. Pliny. *Natural History*, Volume III, Book 11. Cambridge, MA:
      Harvard University Press: 1.
34   Ogilvie, Brian W. 2006. *The Science of Describing: Natural History in Renaissance Europe*.
      Chicago: University of Chicago Press.
35   For a review of the history of insect studies, see Jorink, *Book of Nature*: 181–256.

contemporary observations of animals, both their own and those of others.[36] Gessner's mid-16th century volumes of animals preceded those of Aldrovandi into print, but Gessner did not include insects in his books, even though he was one of the earliest to add information on insects to the writings of antiquity. Instead, much of his unpublished work on insects was obtained by Thomas Penny (English, c. 1532–1589), who added his own observations and those of Edward Wotton (English, 1492–1555).

While it is natural to be curious about who was 'first' to investigate or record some phenomenon, it is an impractical pursuit and one that probably should be abandoned by scholars. On the other hand, it generally is possible to document who was the first to put something into print, although this too can be complicated. Publication dates, even if reliable, do not help us to order the events as they occurred, because as we shall see, the production of a book was often delayed for years or decades after the work was completed. What we do know is that insects were on the minds of Penny, Moffet and Aldrovandi at roughly the same time that Joris Hoefnagel was making his exquisitely detailed paintings of these small animals. Penny's work was continued after his death by his young friend, the English physician Thomas Moffet (1553–1604). Moffet's studies were supplemented with contributions from a number of other people, including John White, Joachim Camerarius the Younger and Carolus Clusius. Moffet did not live to publish his work either, and it ultimately was shepherded into print in 1634 as *Insectorum sive Minimorum Animalium Theatrum* by French physician Théodore Turquet de Mayerne (1573–1655).[37] Emblematic associations of various types of insects appear along with other material from classical authors, but fresh information is added to this mixture. The text includes physical descriptions of insects and some habits and food choices are described. An attempt at ordering insects is made; for example, caterpillars are divided into smooth or hairy groups. The images were printed from woodcuts, and the models for these came from a range of sources compiled over decades.

36   For example, Gessner, *Historia Animalium* and Aldrovandi, Ulisse. 1599. *Vlyssis Aldrovandi ... Ornithologiae, hoc est de Avibus Historiae, Libri XII.* Bononiae: Franciscum de Franciscis Senensem.

37   Moffet et al., 1634. *Insectorum sive Minimorum Animalium Theatrum.* Roughly two-thirds of the content of the finished book concerns what we think of today as insects. The remainder concerns arachnids and worms, which were grouped with insects at the time. Janice Neri provided an excellent reconstruction of the complicated history of this book's genesis and traced the source of some of the illustrations. Mayerne could have been aware of the manuscript through medical interests shared with the older naturalist. Neri, *Insect and Image*: 45–63. See also Vignau-Wilberg, *Archetypa*: 40, regarding the Hoefnagel images used in Moffet.

Some of the images in the *Insectorum Theatrum* were original to the work, others were copied from Hoefnagel's *Archetypa*, work by John White, or from the insect book published by Aldrovandi 32 years earlier.

Written somewhat later than *Insectorum Theatrum* but published earlier, Aldrovandi's 1602 *De Animalibus Insectis Libri VII* was one of the earliest printed books to focus on the natural history of insects. Aldrovandi echoed Pliny as he encouraged the study of the small, writing in his dedication that the smallest creatures exemplify God's wisdom better than do large animals.[38] Like Moffet, Aldrovandi's book combined direct observation and received information dating back to classical authors, but he lived to complete the project and thus controlled the flow of the information appearing in the finished work. Whereas *Insectorum Theatrum* was the output of a number of natural philosophers from different times and countries, Aldrovandi's insect book was under more of his direct control. He hired assistants and artists, and described in his preface how they would accompany him into the field and make notes and drawings as he recounted his observations to them.[39] He also collected specimens, many of which still exist today in Bologna. As in his other animal encyclopedias, Aldrovandi included information from classical scholars, but he added greatly to it. His *De animalibus Insectis* included a dichotomous key to major groups of insects, which he developed based on their habitat (e.g. aquatic vs. terrestrial) as well as external form. The insect groups were designated by their historical Latin names, *Papilio* for lepidopterans, *Formica* for ants and so on, but as in the case of Moffet's work, individual types of insects (what we would call species) largely remained nameless. Some insects, like the noble bee, were given a good deal of attention by Aldrovandi, whereas others were pictured and described more concisely.[40] The images made for him by artists such as Jacobo Ligozzi (Italian, 1547–1627) and Cornelius Schwindt (German, 1566–1632) were compiled into paper museums that Aldrovandi later reorganized according to his classification schemes and integrated with the text of his book. Schwindt was responsible for most of the insects that appeared in Aldrovandi's book, painting them from live or preserved specimens

---

38   Aldrovandi, Ulisse, Gio Battista Bellagamba, Giovanni Luigi Valesio Battista, and Francesco Maria Della Rovere. 1602. *De Animalibus Insectis Libri septem: Cum singulorum Iconibus ad Vivum expressis. Libri VII. Bonon: Apud Ioan: Bapt: Bellagambam cum consensu superiorum*: Dedication.

39   Ibid.: preface.

40   For more on the content of this encyclopedia as well as discussion of contemporaneous insect studies see Ogilvie, Brian W. 2008. Nature's Bible: insects in seventeenth century European art and science. *Tidsskrift for Kulturforskning*, 7(3): 5–21.

and then carving woodblocks for printing.[41] The images of insects created by Schwindt, just as the *Papilio machaon* painted by Joris Hoefnagel, had a reach far beyond their initial publication. The images of the silkworm's metamorphosis in Aldrovandi's *De Animalibus Insectis* (Figure 16) were copied in altered form and used in Moffet's *Insectorum Theatrum* (Figure 17).

Half a century after the appearance of Aldrovandi's insect book, Polish physician and scholar Jan Jonston (1603–1676) published his 1653 volume on insects, which takes its contents largely from Aldrovandi and Moffet's books.[42] The silkworm images originally created by Schwindt made yet another appearance in an edition of Jonston's *Historia Naturalis de Insectis* (Figure 18), published by the Merian family publishing firm. Although Matthaeus the Younger and Caspar, half-brothers to Maria Sibylla, made engravings that could have elevated the level of detail over what could be achieved with woodblock prints, it is not possible to add accurate structural information to an image copied from another. Merely reproducing the work of others eliminated the need for close observation of nature; hence, Jonston's encyclopedia added little new information about insects to what had been published before. However, the Latin edition of Jonston, along with a Dutch edition issued in 1660,[43] helped further to popularize the study of insects, as did Edward Topsell's version of Moffet's work, which was published in English.[44]

Copying of images both with and without attribution was rampant in the creation of all manner of early natural history books. The practice saved time, because no specimens needed to be collected to serve as models for image. Perhaps more important to book production, the cost was reduced, because no draftsperson was needed. The authority of the author and the legitimacy of the information in a book's illustrations could be compromised by the source of the image. If an organism was drawn 'from life' this was usually touted, but the phrase could mean that image source was a dried or preserved specimen. For those who did not have access to a plant or animal in a three-dimensional form of any kind, it was not unusual to copy another image, preferably from a 'trusted' source. Gessner validated his images by tracing his sources, and other

---

41   Neri, *Insect and Image*: 32–45. Neri described this process of 'cutting and pasting' in detail.
42   Jonston et al., *Historiae Naturalis de Insectis*.
43   Jonston, Jan and Matthias Graus. 1660. *I. Jonstons naeukeurige Beschryving van de Natuur der vier-voetige Dieren, Vissen en bloedlooze Waterdieren, Vogelen, Kronkeldieren, Slangen en Draken*. Amsterdam: Jan Jacobsz Schipper.
44   Topsell, Edward, Conrad Gessner, Thomas Moffet, and John Rowland, M.D. 1658. *The History of Four-footed Beasts and Serpents ...* London: Printed by E. Cotes, for G. Sawbridge. Insects are omitted from the title of this book, but the second volume is an English translation of *Insectorum Theatrum*.

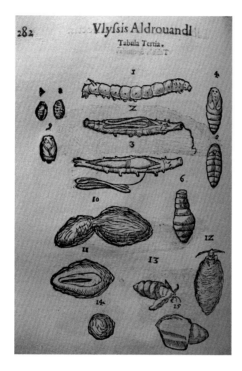

FIGURE 16
Silkworm metamorphosis in Aldrovandi,
1602. *De Animalibus Insectis Libri* VII: 282

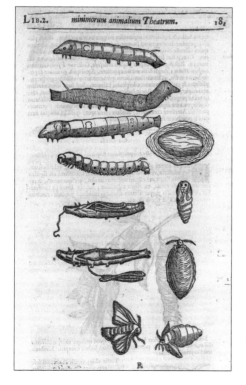

FIGURE 17
Silkworms in Moffet, 1634. *Insectorum Theatrum*, 18

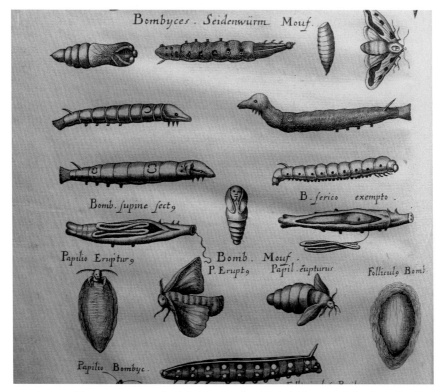

FIGURE 18   Silkworm metamorphosis in Jonston, 1653. *Historia Naturalis Animalium*. Detail
from Plate 22. These images were copied from Moffet.

naturalists attested to the authority of their illustrations in a similar way or by
asserting that they were made *ad vivum*.[45] The spread of printed books naturally caused a proliferation of copying, because source images were increasingly available. Often, a source image was itself a copy, and any errors in earlier
versions were perpetuated. Natural history books were often produced by 'cutting and pasting' from earlier volumes, and information originally gleaned by
others was redistributed, whether accurate or not.

45   Kusukawa, Sachiko. 2010. The sources of Gessner's pictures for the *Historia animalium*.
      Annals of Science, 67(3): 303–328. "From life" had numerous meanings in early natu-
      ral history works. For discussion of this complex subject, see Kusukawa, Sachiko. 2019.
      Ad vivum: Images and knowledge of nature in Early Modern Europe. In *Ad vivum?* Edited
      by Thomas Wolf, Joanna Woodall, and Claus Zittel. Leiden: Brill: 89–211.

Publishing Illustrated Natural History Books

Although less costly than manuscripts, printed books that were heavily illustrated were beyond the means of all but the wealthy through the 17th century. The publishing industry evolved as the demand for books increased. Publications such as herbals were used by medical practitioners as well as collectors, and interest in natural theology helped to drive the market for other natural history books. Collectors and avid gardeners used natural history books much like catalogs for adding to their array of plants or animals and as sources of identification for specimens they had already obtained. However, as discussed above, many if not most naturalists did not publish their findings. As exemplified by the *Insectorum Theatrum*, and Gessner's animal encyclopedias, it was not uncommon for books to be published by associates or followers well after the death of the initial authors or the original creators. Sometimes personal circumstances or the time needed to finish a book defeated a potential author, but often the major obstacle was publishing costs. The cost of publishing a heavily illustrated book in Merian's time depended in large part on the price of paper, which varied with size and quality, and the funds needed to finance production of the plates. Analysis of the economics of printing illustrated nature and medical books in the 16th century reveals that the printed plates could constitute fifty to seventy-five percent of the production costs. Paper was the second highest expense, and the printing of text usually made up less than ten percent of the total.[46] The high price of illustrations is understandable when one considers that their production could involve up to four or more levels of work: the original observers, the draftspeople, the artisans who made the woodblock or copper plate for printing, and the printers.[47] Professional engravers in the 17th century could be paid as much for one detailed copper plate as a skilled mason would earn for nearly a month's labor.[48] If a colorist was added to the list of artisans needed, book costs could double or even

---

46 Kusukawa, Sachiko. 2012. *Picturing the Book of Nature: Image, Text, and Argument in Sixteenth-Century Human Anatomy and Medical Botany.* Chicago: University of Chicago Press: 50–54. These calculations do not include any payment for the author of the work.

47 Images generated with woodblocks can appear on the same page with type and use the same press (see Figure 2). Copper plate printing requires a different type of press and a separate sheet of paper from the letterpress printing used to generate the body of text. Copper plate printing also necessitates the use of more paper, because only one side of the sheet can be printed. Any added text must be etched or engraved onto the plate, a specialized skill that further increases cost.

48 Bowen, Karen L. and Dirk Imhof. 2003. Reputation and wage: The case of engravers who worked for the Plantin Moretus Press. *Simiolus: Netherlands Quarterly for the History of Art 30*: 191.

quadruple. Hence most books were sold without this luxurious addition and were colored post-purchase if at all.[49] Books were usually sold as loose sheets, so the purchasers also bore the cost of having their books bound.

When naturalists were not able to produce their own plates, the high cost of hiring skilled artisans could make publication of an illustrated version of their work prohibitively expensive.[50] Sponsorship and patrons were often the key to the publishing success of such books. Robert Hooke's *Micrographia* was subsidized by the Royal Society and might not have been published without this support. John Ray was unable to receive such sponsorship for his history of insects, a volume that combined his work with that of Francis Willughby (1635–1672).[51] Neither man lived to see their work turned into a book. Ray incorporated his friend's work into a manuscript with his own studies of insects, and this was published in 1710, five years after his own death and without the illustrations he greatly desired to include. In the last five months of his life, Ray wrote to Hans Sloane at least twice seeking help in getting funds for the images, writing in one letter that without illustrations, the book would be a "blind and useless work."[52] He was unable to raise the funds necessary to employ an engraver, and the epilog of his book exhorts its readers to refer to Merian's European and Surinam books for "elegantly depicted and engraved" images of the insects referred to in the text of *Historia Insectorum*. Whether Ray was personally familiar with Merian's work is unclear, although it seems likely he would have known the earlier works on European insects if not her 1705

49   Kusukawa, *Picturing the Book of Nature*: 76.

50   The labor alone for the drawings and engravings needed to create twelve plates in Johann Philip Breyne's 1732 book on fossils was more than half the cost of producing the book, which totaled 482 guilders. The book sold for 3 guilders, requiring sales of 160 copies to cover the production cost. Naturalists who could not afford to finance the publication of their own works faced a loss of control to whoever was footing the bill. Margócsy, Daniel. 2014. *Commercial Visions: Science, Trade, and Visual Culture in the Dutch Golden Age.* Chicago: University of Chicago Press: 80–85.

51   Ray, John. 1710. *Historia insectorum opus posthumum jussu Regiae Societatis Londinensis editum.* London: Impensis A. & J. Churchill. For more on the incomplete nature of the *Historia*, its publication, and the omission of Willughby from authorship see Ogilvie, Brian W. 2016. Willughby on insects. In *Virtuoso by Nature: The Scientific Worlds of Francis Willughby FRS (1635–1672)*. Edited by Tim Birkhead. Emergence of Natural History, Volume 1. Leiden: Brill: 335–59.

52   In a letter to Hans Sloane dated 9 August 1704 Ray proposed to sell subscriptions in advance of publication of his book on insects. Ray, John, William Derham, and E. Lankester. 1848. *The Correspondence of John Ray.* London: Printed for the Royal Society: 448–49.

work on Surinamese insects.[53] The epilogue may have been added by William Derham, who edited the posthumous publication of Ray's insect book.[54]

Studies of Insect Metamorphosis

In the early days of insect studies, specialization on one group of animals was rare, and few descriptions of insect metamorphosis were published. Natural philosophers like Aldrovandi and Gessner systematically surveyed broad groups of organisms, and insects were a relatively minor subject for many of these scholars. Like these early encyclopedists, many naturalists had medical training that guided their studies; for example, Thomas Moffet was interested in nutrition and wrote about Hippocrates. In his relatively short life, Willughby investigated birds and fish in addition to insects, and Ray devoted much of his time to plants. Through the first half of the 17th century, much of the effort devoted to insects as well as other organisms was focused on ordering and classification rather than details of form and function. The insect encyclopedias contained descriptions of external form, attempts at classification and some information on habits and reproduction, but there was no sustained or in-depth study of these topics by the aforementioned naturalists. However, interest in insects fueled by natural theology's admonitions to find God in the small, and the availability of the early 17th century encyclopedias converged in a time when direct observations of nature were becoming of more interest than the wisdom of the ancients. Emblematic nature faded into the background as empirical studies blossomed, aided by improvements in magnification techniques and spurred by questions about the internal processes of organisms. By the second half of the 17th century, insects were at the center of questions about spontaneous generation, reproduction and metamorphosis.

The manner in which Aldrovandi, Moffet and their imitators addressed metamorphosis in insects seems puzzling today. Although they understood that the caterpillars would become pupae and then winged insects, their books did not visually connect larvae with specific adults. In the insect encyclopedias described above, the material on adult lepidopterans, although cross-referenced in some places, typically was separated by several pages from the images and text related to their larval stages (see Figure 19 for a later example

53   Merian, *Metamorphosis*.
54   Ray, *Historia insectorum*: 398. The epilog consists of a short note suggesting that readers consult images in Merian's two quarto volumes (on European insects) and her 1705 *Metamorphosis* (on insects in Surinam).

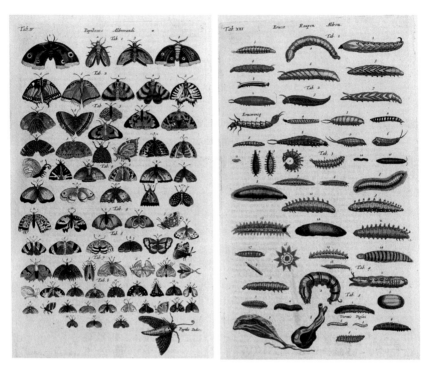

FIGURE 19   Plates IV and XXI from Jonston's book on insects were originally engraved by
              Matthaeus Merian Jr. and Caspar Merian for the Latin edition (Jonston, 1653).
              This image is from a hand-colored Dutch edition (Jonston, 1660).

of this visual segregation). One exception to this was the silkworm moth (see
for example Figures 16, 17 and 18), which had been bred in captivity for centuries. Of more interest, Moffet or his associates included information about
the metamorphosis of a native European species of moth in their work. In this
marked exception to the separation of stages, a relatively accurate if crude
portrayal of the eggs, caterpillars of two sizes, pupa and adult of what is now
known as the garden tiger moth (*Arctia caja*) were shown (Figure 20). This and
the brief physical description of the stages indicate that at least one of the con-
tributors to the Moffet volume observed and recorded this transformation.[55]
English scholar Francis Willughby studied insects and investigated aspects
of their metamorphosis, but like his contemporaries, he grouped larvae and
adults separately. His friend John Ray also raised a variety of caterpillars
through metamorphosis to determine which adults would emerge from which
pupae. Both men were more interested in using the differing ways that insects

---

55   Moffet, *Insectorum Theatrum*: 186.

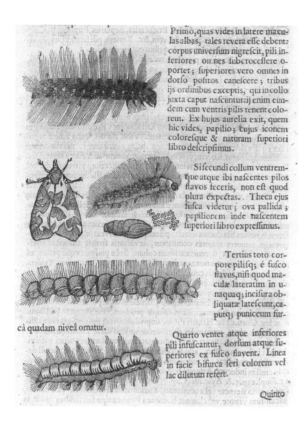

FIGURE 20
Arctia caja (garden tiger moth) life stages from eggs to adult moth. Moffet et al., 1634. *Insectorum Theatrum*, 186

transform to classify them into groups than in metamorphosis as a process. In addition to their efforts at creating a systematic ordering of insects, Willughby and Ray observed some insect behaviors and noted the presence of parasitoids in some groups.[56] Their work added to what was published by their predecessors, but most of their findings were not printed until 1710, decades after much more extensive descriptions of insect metamorphosis and behavior were published by Johannes Goedaert and Maria Sibylla Merian.[57]

By the middle of the 17th century, new questions came to the forefront as the development of improved magnification techniques facilitated tremendous strides in understanding of insect structure and internal anatomy. Robert

---

56   A letter from Willughby to the Royal Society of London concerned parasitoids, in this case a type of wasp that frequently kills caterpillars and pupae. The term parasitoid designates a parasite that kills its host. Willughby, Francis, 1671. "A letter of Francis Willoughby Esquire, of August 24, 1671. Containing some considerable observations about that kind of wasps, call'd Vespæ Ichneumones; especially their several ways of breeding, and among them, that odd way of laying their eggs in the bodies of caterpillars, &c." *Philosophical Transactions of the Royal Society of London* 6(76): 2279–281.

57   Ray, *Historia Insectorum*.

Hooke's 1665 *Micrographia* broke new visual ground with his stunning images of flies, gnats, and other insects, but he was very broad in his research interests, most of which did not involve natural history.[58] Around the same time that Hooke produced his masterpiece, Marcello Malpighi (Italian, 1628–1694), Antonie van Leeuwenhoek (Dutch, 1632–1723), and Johannes Swammerdam (Dutch, 1637–1680) were investigating processes in insects that could be illuminated by dissection and microscopic examination, i.e. digestion, excretion, reproduction, development, and growth. Swammerdam made observations of insect metamorphosis for a limited number of species and published some of these in his 1669 *Historia Generalis Insectorum*.[59] Both he and Francesco Redi (Italian, 1626–1697) provided evidence to refute long held belief in the spontaneous generation of insects. For Swammerdam it was imperative to demonstrate that insects had internal structures, and that they grew and developed like all other animals, because this was evidence that God had the same rules for all creatures. Swammerdam meant for his *Historia* to encourage piety, and to show that because God was wondrous, there was no room for random events in his creations.[60] However, like Redi, whose studies ranged beyond insects to venoms and parasites, Swammerdam had broader interests apart from insects, and he also examined metamorphosis in frogs and the structure of muscles.

In roughly the same period that the above-mentioned microscopists were probing the inner workings of insects, equally significant work was performed by naturalists making larger scale observations in the field and laboratory. Malpighi's carefully rendered details of silkworm anatomy as seen under magnification present very different kinds of information from what is conveyed by Maria Sibylla Merian's depiction of the stages of the silkworm's life cycle (Figure 21 here and Merian's Plate 1 on p. 148 of this book).[61] Although Merian occasionally used magnification and conducted simple dissections, investigations into the ultrastructure of insects did not influence her early work to the degree that Johannes Goedaert's studies on metamorphosis did. He preceded

---

58   Hooke, Robert. 1665. *Micrographia, or, Some Physiological Descriptions of Minute Bodies Made by Magnifying Glasses with Observations and Inquiries Thereupon* / by R. Hooke ... London: Printed by Jo. Martyn and James Allestry, printers to the Royal Society. For more on the insects in Hooke's work, see Neri, Janice. 2008. Between observation and image: representations of insects in Robert Hooke's Micrographia. In *Art of Natural History*. Edited by Theresa O'Malley and Amy Meyers. Washington DC: National Gallery of Art: 83–107.

59   Swammerdam, Johannes. 1669. *Historia Generalis Insectorum, ofte Algemeene Verhandeling van de Bloedeloose Dierkens*. Utrecht: M. van Dreunen.

60   Jorink, *Reading the Book of Nature*: 226–28.

61   Malpighi, Marcello. 1669. *Marcelli Malpighii Philosophi & Medici Bononiensis Dissertatio Epistolica de Bombyce*. London: Joannem Martyn & Jacobum Allestry.

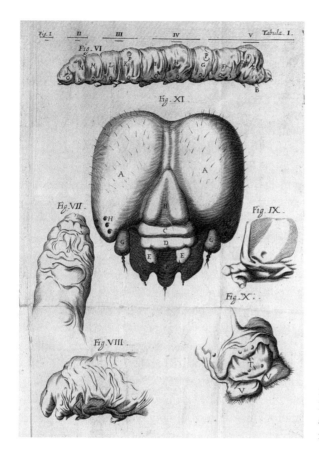

FIGURE 21
Details of silkworm
anatomy from Malpighi,
1669. *De Bombyce*, Plate 1

her by a few decades in becoming the earliest naturalist to specialize in the investigation of insects, beginning his studies around 1635. He lived his entire life from 1617 to 1668 in Middelburg, training in art and then becoming a still-life and landscape painter with an interest in alchemy and an even deeper interest in religion. When he turned his attention to insects, he seems to have found a new passion, for he investigated more than 100 species, recording their habits and metamorphoses. Many of his insect subjects were moths and butterflies, but he also included entries on woodlice, arachnids and tiny crustaceans. He was driven to do this work to show that "none of God's creatures is to be despised ..." and to bring the unexplored (insects) "into the light" for the glory of God.[62]

The composition of Goedaert's plates in his *Metamorphosis Naturalis* were a step beyond the images in the earlier insect encyclopedias, because he

---

62   Goedaert, *Metamorphosis Naturalis*, Volume 1: Dedication.

combined the life stages of each type of insect on a page (Figures 3 and 22). Earlier naturalists could not be sure of the associations between larval, pupal and adult forms, because the only definitive way to make these connections is to capture and raise them through all stages, a labor intensive process requiring dedicated attention to insects. Aside from presenting the metamorphic stages of each species together and including the parasitoids that prey upon some insects, Goedaert's images did not differ dramatically from what came before. He isolated each insect and displayed it in a way such that its physical traits could be examined, static compositions that reveals little about the life of the animals. He was careful to make his insects life-sized, something not always achieved by his predecessors; for a large moth or caterpillar he employed a fold-out page to accommodate its size. The arrangement of insect life stages in his books are usually chronological, echoing the scheme used in the watercolors of Alexander Marshal, to whom this sequence also made sense (Figure 10 above). Goedaert and in some instances, Aldrovandi and Moffet, occasionally included an image of a caterpillar on a leaf but this was the exception rather than the rule. Plant hosts usually were not pictured in the early insect books, nor were they discussed in detail or with much accuracy.

Goedaert makes it clear that his findings were the result of first-hand observations, and that in pursuit of this, he captured and raised insects in glass jars. Like all naturalists and indeed scientists today, Goedaert made mistakes. For example, he believed that insects arose by spontaneous generation, and he describes this process throughout his volumes, although in seeming contradiction, he describes and pictures eggs for a few species. Both Aristotle and Pliny had put their stamp on the idea of spontaneous generation of insects, an idea that was to remain in place until Redi, Swammerdam, Merian and others rejected it in the 17th century.[63] And unlike Merian's plates, several of Goedaert's are missing at least one metamorphic stage (Figure 22). He did not always know the food required by a species. In some cases he lost larvae or pupae to parasitoids and so was unable to get all stages to thrive. However, he accurately recorded many aspects of insect behavior and life histories, some of which were echoed by Merian's writings. Of significance, too, is the fact that Goedaert published his first edition in Dutch, an unusual choice at a time when most scholarly books were written in Latin, and thus accessible only to the educated reader. It may be that he used the vernacular because he did not have abilities

---

63   Belief in spontaneous generation for some organisms maintained a tenacious foothold well into the 18th century. See for example Terrall, Mary. 2018. Experimental natural history. In *Worlds of Natural History*. Edited by Helen Anne Curry, Nicholas Jardine, James A. Secord and Emma C. Spary. Cambridge: Cambridge University Press: 178–80.

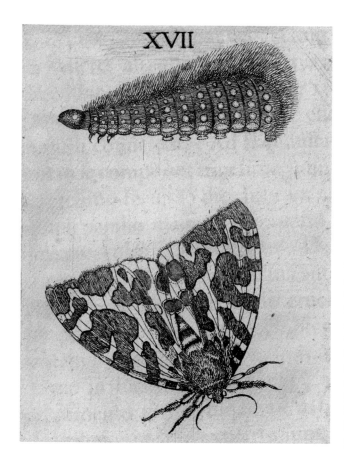

FIGURE 22
Arctia caja (garden
tiger moth) in
Goedaert, 1660.
*Metamorphosis
Naturalis*, Volume
1, Plate 17 . The
complete life cycle
of this moth was
pictured by Merian
in Plate 5 of her 1679
Raupen book (see
p. 168 in this book).

in Latin, but the end result made his work accessible to an additional audience. Eventual translation into Latin by Johannes de Mey broadened the audience for Goedaert's books, as many more scholars at the time read Latin than could read Dutch.

True to his purpose for the work, Goedaert's books have a religious message for his readers. Like many Christians, he saw insects such as the butterfly as a perfect symbol of the resurrection of Christ. In his view the caterpillar "died" and the new form arose from the old, providing a message of hope. Goedaert also used the example of a moth drawn fatally to the light of a flame as an allegorical warning for those curious about divine secrets, who could be "swallowed up by the radiance" of God's majesty in their search.[64] Goedaert's careful observations were colored by his piety, and they were supplemented with commentary and scholarly references from a Middelburg pastor, his friend Johannes de Mey.

---

64   Goedaert, 1660. *Metamorphosis Naturalis*: 18.

As well as his contribution to the translation of the books into Latin, De Mey provided much of the impetus behind the writing of *Metamorphosis Naturalis*, and he worked diligently to popularize the three volumes.[65] The images and text of Goedaert's work received further exposure and analysis in the hands of English naturalist Martin Lister (1639–1712), who translated Goedaert into English and reorganized the existing Latin translations.[66] In another example of the obstacles of expense, publication of Lister's translation of Goedaert was delayed almost a decade by the cost of printing the work, even though fifty copies were funded in advance by the Royal Society of London.[67] Lister completely reordered his entries according to his own taxonomic ideas, and had new plates made using copies of Goedaert's images to reflect this organization. His commentary is critical of Goedaert in several places, but Lister was working with the benefit of enhanced knowledge of insects from his own studies and those of others, such as studies of lepidopteran parasitoids by Willughby. Lister particularly took Goedaert to task for neglecting to include the specifics of the preferred larval food plants, writing that it is necessary that a naturalist be skilled in plants. He also complained about another deficiency: "I say again I could have wished the author had taken the care of describing to us the paintings and color of the insects...."[68]

Maria Sibylla Merian was raising insects through metamorphosis at roughly the same time that Goedaert's first volume was published, and after his work and hers became well known, many naturalists began to follow suit, aided by the information that these entomological pioneers provided.[69] Knowing which larva would become a certain adult and what foods were necessary to

---

65   Jorink, *Reading the Book of Nature*: 206. De Mey was assisted in the translation by Paulus Veezaerdt.

66   Goedaert, Johannes and Martin Lister. 1682. *Johannes Goedart of Insects*. York: printed by John White for Martin Lister and Goedaert, Johannes and Martin Lister. 1685. *Johannes Goedartius de Insectis, in Methodum Redactus, cum Notularum Additione. Operâ M. Lister.* London: S. Smith.

67   Roos, Anna Marie. 2018. *Martin Lister and his Remarkable Daughters: Art and Science in the Seventeenth Century.* Oxford: Bodleian Library Press: 91. Conchology is the study of shelled mollusks. Dismayed by the cost of having prints made for his version of Goedaert's work, Lister trained his young daughters in etching in order to reduce the cost of illustrating his later work on conchology.

68   Goedaert and Lister, *Of Insects*: 41. Lister bemoans the fact that "the author has left us in the dark for the food of this insect" for want of a particular name for the plant on which it feeds. On p. 48 he refers to Goedaert's lack of color descriptions for the insects.

69   Merian began in 1660 as a young woman, and she discovered Goedaert's work later (see p. 89 in this book).

make this possible would have saved other aspiring entomologists a great deal of time and effort.

In addition to their specialization in insects, Goedaert and Merian were unusual for their time in that they recorded their own observations in word and image, producing their books' plates themselves. This cost savings made timely publication feasible and avoided the effects of using intermediary artisans for drawing and etching. However, Merian's books differed from Goedaert's in several ways. Although she was familiar with his work, her tone was less overtly religious, and she did not apply symbolic meaning to her subjects. She eschewed spontaneous generation as the origin of moths and butterflies, frequently discussing and picturing their eggs. Her work on insects spanned almost twice as many years as did that of Goedaert, and her persistence in seeking the full life cycle of her subjects led to almost 200 complete metamorphic cycles recorded in her European insect books alone, and dozens more in her Surinam volume. Her text made clear the colors of all stages of the insects, aiding greatly in identification. However, the "novel invention" she refers to in the title of her 1679 caterpillar book was what set her work apart from all books on insects that preceded hers. By presenting the metamorphic stages of insects together with their larval food plant, she made a fundamental connection evident to her readers. Furthermore, Merian's ability to render images packed with information was well suited to this groundbreaking composition. The structural details found in earlier images of insects were still possible in her plates, but by including a plant in the frame, she also could portray insects as flying, crawling, and laying eggs. In her *Raupen* books, caterpillars feed on plants and are fed upon by parasitoids, and in some images, adult lepidopterans extract the nectar of flowers. Her plates illustrate other ways that insects interact with the plants, attaching silk threads, rolling the leaves as a shelter, living in a plant gall, or burrowing through a rose bud. Action had entered the area of natural history illustration at the level of some of the smallest and most common animals, and their integral relationships with plants were emphasized. These additions to the knowledge base on insects were made possible by her skills as an observant naturalist, artist and printmaker, capable of creating high quality, accurate plates. Merian fit the ideal model of image-maker as envisioned by Carl Linnaeus, who stated that the best images of the plants he was attempting to classify were made by botanists who themselves were skilled at drawing and engraving.[70] The direct evidence of the influence of

---

70   Linnaeus, Carl. 1751. *Philosophia Botanica in qua explicantur Fundamenta Botanica cum Definitionibus Partium, Exemplis Terminorum, Observationibus Rariorum, adjectis Figuris Aeneis.* Stockholm: Godofr. Kiesewette: 263. Neri discussed this partnership of naturalist

Merian's images as shown by the epilogue in Ray's insect history is telling. The equally important information in her text complements her images. How she came to produce her work on insects and the significance of her investigations will be addressed in the following chapters.

and artists, observing that the former provides the reliability or validity for the image and reigns in the creative impulses of the artist. Neri, *Insect and Image*, xix. For more on collaboration between naturalists and artisans see Nickelsen, Kärin. 2018. Images and nature. In *Worlds of Natural History*, Editors: Helen Anne Curry, Nicholas Jardine, James A. Secord and Emma C. Spary. Cambridge: Cambridge University Press: 229–32.

CHAPTER 2

A Life Investigated

Maria Sibylla's personal story has been explored by many, yet for much of her life we know little about her specific activities. She would probably be the first to agree that the core of her life could be viewed and best understood through the lens of her work. Very little remains in the way of personal writings; for example, only eighteen letters that she wrote have been discovered. Her biography has been pieced together by a number of scholars using these letters, her study journal, her published works, the writings of her contemporaries, and archival sources from the period in which she lived.[1] Here I examine what is known of her history up through the publication of the 1679 *Raupen* book, focusing on factors that led her to produce this remarkable work at age 32 and to pursue decades of research on insects. Merian was endowed with both artistic talent and natural curiosity, and she was born into a fortunate situation for exercising these attributes. Critical influences ranged from her early family life and training in Frankfurt, to her formation of personal networks in Nuremberg. The fact that a woman in 17th century Germany produced pivotal natural history books is somewhat less surprising when one delves into the milieu of Frankfurt am Main, Merian's home until she was 18, and Nuremberg, where she conducted much of the work for her first two caterpillar books, published in 1679 and 1683.[2] Both cities were extraordinarily rich in the visual and intellectual resources needed for such an endeavor, and societal norms fostered creative industry among both women and men.

Growing up in Frankfurt am Main

Johannes Gutenberg developed his moveable type press in Mainz, less than 50 km from Frankfurt, so it is not surprising that the latter city was the center

---

1 For more on Merian's family, marriage and life after 1680 see for example Davis, Natalie Zemon. 1997. *Women on the Margins: Three Seventeenth-Century Lives*. Cambridge, MA: Harvard University Press: 140–202; Lölhöffel, Margot. 2015. Maria Sibylla Merianin und Johann Andreas Graff: Gemeinsames und Trennendes, Erster Teil. *Nürnberger Altstadtberichte* 40: 36–76; Todd, Kim. 2007. *Chrysalis: Maria Sibylla Merian and the Secrets of Metamorphosis*. Orlando: Harcourt; and Merian, Maria Sibylla and Kurt Wettengl. 1998. *Maria Sibylla Merian, 1647–1717: Artist and Naturalist*. Edited by Kurt Wettengl. Ostfildern-Ruit: Verlag Gerd Hatje.
2 Merian, 1679 and 1683. *Raupen*.

© KONINKLIJKE BRILL NV, LEIDEN, 2021 | DOI:10.1163/9789004284807_003

FIGURE 23   Detail from engraving of Frankfurt on the Main: View of the city as seen from
               the southwest. Matthaeus Merian the Elder. Circa 1617

of international book trade. Frankfurt was home to the largest book fair in
Europe after the fair was established around 1475. The Main river, a large tribu-
tary of the Rhine, formed a ready transportation route, enhancing Frankfurt's
location astride major trade routes connecting European centers from north to
south and east to west (Figure 23).[3]

At the turn of the 16th century, Frankfurt offered a bounty of opportunities
that attracted artists like the Hoefnagels, Georg Flegel, Jacob Marrel, and naturalists such as Carolus Clusius, all of whom moved there from other cities. This
city of books was awash in engravers as well as artists, and many publishers
set up shop in Frankfurt. Their highly varied output included religious works,
emblem books, florilegia and influential herbals such as those by Mattioli.[4]
The very successful firm of Theodor de Bry specialized in engravings for illustrated books and eventually moved into publishing and bookselling. The De
Bry firm was particularly known for its books on exploration of the Americas,
presenting images of plants and animals as well as native peoples, but they
also published a number of works featuring nature and medicine.[5] Matthaeus
Merian the Elder (1593–1650) moved from Basel and joined De Bry in 1616, and
a year later, he married the daughter of the house, Johanna Magdalena de Bry.
He and their sons, Matthaeus the Younger and Caspar, eventually took over

---

3   Pettegree, Andrew. 2007. *The French Book and the European Book World.* Leiden: Brill: 129–32.
4   Mattioli, Pietro Andrea, Joachim Camerarius, and Francesco Calzolari. 1586. *De Plantis
    Epitome vtilissima.* Frankfurt: Feyerabend.
5   Van Groesen, Michiel. 2008. *The Representations of the Overseas World in the De Bry Collection
    of Voyages (1590–1634).* Leiden: Brill: 80–2, 92 and 139–74.

the business, and it thrived under the Merian name for decades. In 1647, Maria Sibylla was born to the second wife of Matthaeus Merian, Johanna Sibylla Heim (German 1620–1690). Just a year before her birth, peace treaties ended the thirty years of war that had ravaged Europe. The first eighteen years of Maria Sibylla's life were spent in Frankfurt, where the new peace and relative economic stability provided beneficial conditions for both art and science, as well as for activities such as gardening and collecting. At home, she was immersed in an environment of books, prints, art, and the business of book publishing.

Matthaeus Merian the Elder's books and prints, like those produced by his father-in-law's firm, often featured nature, both real and imagined. These images must have left an impression on the young girl and would have contributed greatly to her informal home education. Even though she did not have the opportunities for travel afforded to her male relatives, Maria Sibylla could have learned something of the world from the resources available at home. Her father made a number of engravings showing towns and cities with landscapes and plants incorporated to varying degrees (Figure 24).[6] One engraving that may have been examined closely by his daughter was the view of Heidelberg produced in 1620 (Figure 25). The extensive pleasure gardens of the Palatine prince shown in the upper left were referenced by his daughter in the preface to her 1680 book of flowers (a *Blumen* book), possibly a nod to this panoramic and tranquil image.[7] A florilegium published by Theodor de Bry with plates by Matthaeus Merian was reissued by the family firm as an expanded edition in 1641.[8] In the *Florilegium Renovatum*, insects were incorporated into four plates, additions that may have caught the eye of the young girl as she was learning to paint by copying works at hand. The sumptuous French florilegium by Daniel Rabel, *Theatrum Florae*,[9] was the image source for four of the insects in *Florilegium Renovatum*, which was published almost two decades

6 See for example Merian, Matthaeus. 1646. *Der Fruchtbringenden Gesellschaft Nahmen, Vorhaben, Gemählde und Wörter*. Frankfurt: Merian. Each image is printed above a short poem honoring a member of the society celebrated by this book. The lower banner displays the member's name within the society; the plant reflects the character or actions of the member. The plate in Figure 24 refers to Joachim von Boeselager, and the medicinal properties of the black horehound plant may symbolize his role as a healer or mediator.

7 Merian, Maria Sibylla. 1680. *Neues Blumenbuch*. Nuremberg: Johann Andreas Graff: Preface.

8 Bry, Johann Theodor de and Matthaeus Merian. 1641. *Florilegium Renovatum et Auctum: variorum maximeque rariorum Germinum, Florum ac Plantarum*. Frankfurt: Matthaeus Merian. This type of book was valued by gardeners to identify flowers for their gardens.

9 Rabel, Daniel. 1622. *Theatrum Florae in quo ex toto Orbe Selecti Mirabiles Venustiores ac praecipui Flores tanquam ab ipsus Deae sinu proferuntur Guillelmus Theodorus pinxit*. Paris: N. de Mathonière. In addition to moths and butterflies, Rabel pictured a grasshopper and

FIGURE 24
Engraved plate by Matthaeus
Merian, 1646. From *Der
Fruchtbringenden Gesellschaft
Nahmen*. The plant is referred
to in the accompanying
poem as black horehound,
which has various healing
properties.

later (Figure 26). Rabel's florilegium was unusual for its time in having insects
adorning several plates, including ten different adult moths and butterflies,
most identifiable as to species, and one sphinx moth caterpillar. These books
may have been available for perusal and copying by the young Maria Sibylla.

Likewise, Maria Sibylla's views of God and nature may have been shaped
by works like her father's engraving, which accompanied Robert Fludd's description of the cosmos (Figure 27).[10] The composition of Matthaeus Merian's
image of heaven and earth is reflected in language used in her first work on

---

three distinct types of beetle, including the ubiquitous stag beetle depicted by Dürer,
Hoefnagel, and numerous other artists of the period.

10   Fludd, Robert. 1617. *Utriusque Cosmi Maioris Scilicet et Minoris Metaphysica, Physica atqve
Technica Historia*. Oppenheim: Theodor de Bry. Volume 1: plate after p. 3.

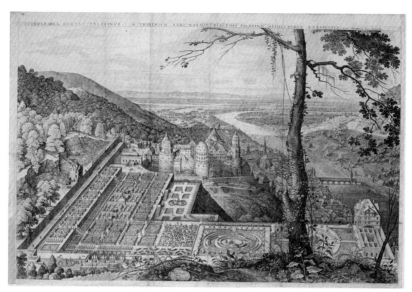

FIGURE 25   Matthaeus Merian, 1620. *Speculum Romanae Magnificentiae Scenographia Hortus Palatinus a Friderico Rege Boemiae electore Palatino Heidelbergae extructus* . This large engraving was printed from four plates on four joined sheets of paper. Frederick V, the Elector Palatine, commissioned the garden design for his Heidelberg castle in 1614, and it was considered one of the finest gardens in Europe.

FIGURE 26   *Aglais io* (peacock butterfly) in Rabel, 1622. *Theatrum Florae*, Plate 39 (left) and a close copy in De Bry and Merian, 1641. *Florilegium Renovatum*, Plate 27 (right)

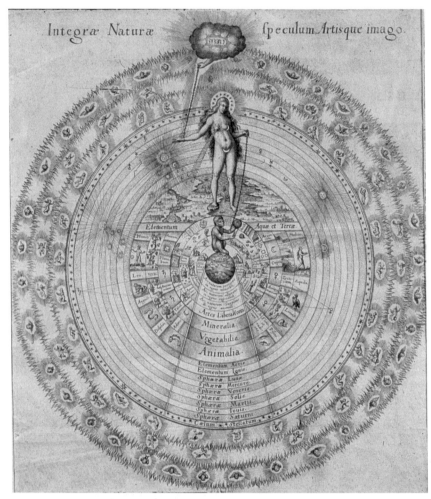

FIGURE 27   Art as the Mirror of All Nature. Engraving of the cosmos by Matthaeus Merian.
                 In Fludd, 1617. *Utriusque Cosmi*

natural history as she refers to God's handmaiden, Nature.[11] Examination of
the concentric spheres that constitute the earthly portion of Fludd's cosmos
reveals that God's handmaiden has given rise to minerals, vegetation and animals, including the smallest creatures, shown here in the form of a snail, bees
and other insects. Man's art in turn 'apes' Nature, the art of God.

Her father's body of work was more influential on Maria Sibylla than his
presence in her life; he died in 1650 when she was three years old. By 1651, her

---

11   Merian, 1679. *Raupen*, Chapter 32.

FIGURE 28   Attributed to Jacob Marrel, 1679. Portrait of 32-year-old Maria Sibylla
              Merian. Oil on canvas

mother had remarried and another artistic influence came into her life. Her
stepfather, the painter Jacob Marrel, would become an important mentor for
the young girl; they appear to have remained close well into her adulthood,
when he may have painted her portrait (Figure 28). Although Maria Sibylla's
mother and new husband would have set up their own household, there is evidence that her half-brother Caspar Merian remained influential in her life, and
it is possible that he was one of her teachers in the arts. Caspar was a talented
printmaker, and she may have learned etching from him. No matter which
family members served as her teachers, we know that she not only learned
to read and write (in German, if not in Latin), but was taught painting and

intaglio techniques. The production of artwork depicting nature, or work in a trade such as publishing, were not the rarities we might imagine for a 17th century German woman. Children of both sexes were often employed in a family business and trained according to individual abilities. Women were not prohibited from working in various trades at this time, particularly if they were part of a family firm; however, they could not join guilds or art academies.

One possible stimulus for Maria Sibylla's childhood insect studies may have been the captivating rows of caterpillars lined up in Jan Jonston's *Historia animalium*,[12] published by the Merian firm when she was six years old (Figure 19 in Chapter 1). Other models available for her study were the paintings by her stepfather. In addition to training with Georg Flegel, Marrel studied under Jan Davidsz de Heem (1606–1684), an influential still-life painter. Marrel passed this training along to his apprentices, Abraham Mignon (Dutch 1640–1679) and Johann Andreas Graff (German 1636–1701) as well as to his stepdaughter. Although Marrel painted the complex still life works in oil that were popular at the time (see Figure 15 in Chapter 1), his portraits of individual tulips and the work of his stepdaughter, more closely resemble Flegel's elegantly simple studies of plants and insects (Figure 29). Indeed, Merian's silkworm study is markedly similar to one by Flegel in several elements (compare Figure 61 to Figure 30 below). Aside from Flegel's silkworm portrayal, his paintings and those of Jacob Marrel did not show the life cycle of moths or butterflies, and the insects in their images had no intentional biological relationship with the plants.

Perhaps inspired by Flegel, Marrel inserted insects into his gouache portraits of individual tulips painted on vellum, dozens of which survive today. In these 17th century tulip portraits, individual flowers were the stars on each page, and the varietal names were usually included for identification. The names, colors and price lists included in some of these specialized volumes could have been used by merchants or served as records of what plant collectors paid for their treasures. The costly materials used for many of these works (e.g., vellum rather than paper as a surface) and the fact that well-established artists made these tulip images, make it probable that many of these works were created for wealthy collectors desiring a permanent record of the glory of their prized tulips.[13] Interestingly, Jacob Marrel's tulip portraits are very different from

---

12   Jonston, Merian and Merian, *Historiae Naturalis de Insectis.*
13   The role of tulip portraits is discussed in Goldgar, Anna. 2008. *Tulipmania: Money, Honor, and Knowledge in the Dutch Golden Age.* Chicago: University of Chicago Press. Passim (e.g., p. 100). At the height of "tulipmania," buyers spent the equivalent of hundreds or even

FIGURE 29
Georg Flegel, c. 1630. Flowers
and caterpillar (probably
Calliteara pudibunda, the pale
tussock moth). Watercolor
and body color on paper

FIGURE 30   Georg Flegel, c. 1630. Silkworm in various stages of
metamorphosis. Watercolor and body color on paper

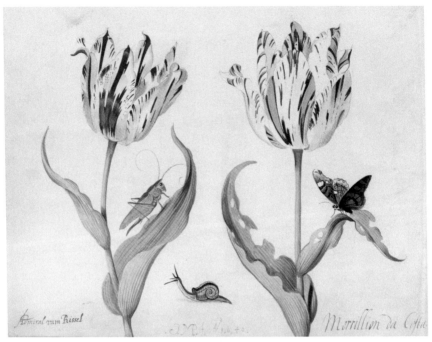

FIGURE 31   Jacob Marrel, 1640. Bodycolor on vellum. Marrel's initials and the year are
             inscribed bottom center.

those by his contemporaries such as Pieter Holsteyn the Younger and Judith
Leyster, in which the tulips are isolated on the sheet and are not associated
with any other organisms. In contrast, Marrel's tulip portraits feature insects
busily conducting their activities in and around the central flowers: bees visit
blossoms, butterflies and damselflies hover, and caterpillars munch their way
across a leaf (Figure 31). Marrel occasionally added the realistic touch of a
chewed leaf, which seems to have been a singular practice among the tulip
book painters of the time. One might suppose that tulip connoisseurs would
desire perfection in these portraits of their treasures. Perhaps the fauna added
by Marrel and their effects on the plants were intended to help collectors to
picture the plants growing in their own gardens. Many of the insects are naturalistic enough to be identifiable to species, and thus were either painted from
specimens he had, or were copied from a reliable image. Marrel's bush cricket

---

thousands of dollars for a single bulb. Jacob Marrel traded in bulbs and could have used
tulip images to help market the more expensive varieties. An inventory of Marrel's goods
made in 1649 included "eight drawers of bulbs" (*Acht tulpen laden ...*). Bredius, A. 1915.
*Künstler-Inventare: Urkunden zur Geschichte der Holländischen Kunst des 16ten, 17ten und
18ten Jahrhunderts*. Den Haag: M. Nijhoff: 114.

(probably *Tettigonia viridissima*) may have been modeled after one in Jacob Hoefnagel's work (Figure 12 in Chapter 1).[14] 'Copy collections' were made for use by Marrel's teacher Georg Flegel and other artists of the era, and this habit was passed on to Maria Sibylla, who worked from notes and studies throughout her life. Marrel reused a variety of small animals in different compositions, and many of the same insects make multiple appearances throughout his tulip portraits.[15]

Whatever Marrel's motivation for these decorative additions to his tulip paintings, a strong case can be made for Maria Sibylla to have taken inspiration from her stepfather's work. One example of Marrel's work is particularly intriguing, because the caterpillar portrayed in Figure 32 is a naturalistic image of one of the Tortricid family larvae, a common pest on roses. Maria Sibylla featured two similar caterpillars in her first *Raupen* book, and she wrote with evident delight about their behavior of climbing up and down the silken thread that is spun from their mouths.[16] This tiny acrobat would have been greatly entertaining to a child, and it could have been shown to her by her stepfather, who painted these caterpillars a number of times.

Early Fascination with Insects

Like most artists of the period, Merian started her training in art by copying the work of others. However, she eventually went on to paint from nature, and for this she began to secure her own subjects, capturing and raising insects on her own. She documented her first independent insect studies as beginning in 1660, when she was thirteen.[17] In addition to sparking her interest in insects with his paintings, Marrel may have introduced her to the metamorphosis of silkworms. Frankfurt was a center of silk production and his brother was

---

14   Hoefnagel, Joris and Jacob Hoefnagel, *Archetypa Studiaque Patris.*
15   I have examined four bound sets of paintings attributed to Jacob Marrel in the Oak Spring Garden Foundation collection and one bound set in the Rijksmuseum, Amsterdam. With the exception of one manuscript at Oak Spring, all paintings are body color on vellum.
16   Merian, 1679. *Raupen*, Chapters 19 and 24. See p. 228 in this volume.
17   See Appendix, study journal entry 1. Her handwritten manuscript and study images of metamorphosing insects is in the collection of the Library of the Academy of Sciences in St. Petersburg, Russia. Throughout this book, I refer to this as her study journal, but it was in essence a research notebook. Entries begin with a description and images of silkworm metamorphosis and continue for her European insects and then with her insects from Surinam. My source is a facsimile edition of this manuscript: Merian, Maria Sibylla and Wolfgang Dietrich Beer. 1976. *Maria Sibylla Merian: Schmetterlinge, Käfer und andere Insekten. Leningrader Studienbuch* ["Book of Notes and Studies"]. Luzern: Reich.

FIGURE 32   Jacob Marrel, c. 1634. Body color on vellum

involved in this trade,[18] perhaps providing her with insects to study, or specimens to paint. Decades later, she harked back to her early experiences:

> From my youth onward I have pursued the study of insects. Initially I
> began with the silkworms in my home town of Frankfurt am Main. Later I
> noticed that far lovelier butterflies and moths emerged from other cater-
> pillars than from silkworms. This moved me to gather all the caterpillars
> that I could find in order to observe their transformation.[19]

Although it is clear that Merian's interest in insects dates from an early age, the
exact sequence and timing of her original studies are difficult to document,
because we do not have the original versions of her early notes. As will be described in the next chapter of this volume, the manuscript of her study journal
that exists today is a compilation of handwritten notes and painted studies
that she recopied and began to compile between 1685 and 1691, when she re-
sided in Friesland.[20] This study journal contains her research notes but also
provides glimpses into parts of her life, insofar as she remarks on where she
collected her insects and in some cases, when. She begins the journal with a
description of silkworm metamorphosis and explains that this insect was presented first, both in her study notes and in her 1679 *Raupen* book, because it is
the most useful and noblest of the caterpillars.[21] While she may have observed
and painted silkworm caterpillars and moths that were raised by commercial
silk producers, her handwritten study notes report that her first independent
experiment with metamorphosis involved the native tortoiseshell caterpillar.
In the Nuremberg summers, this species would have been locally abundant,
feeding on the ubiquitous nettle plant.[22] Whether she continued her insect
studies from age 13 until her marriage five years later is not known, but by early
adulthood, she was an accomplished artist and a budding naturalist.

18   Davis, *Women on the Margins*: 146.
19   Merian, *Metamorphosis*, preface. All translations of *Metamorphosis* into English have
      been taken from Merian, Maria Sibylla, Marieke van Delft, and Hans Mulder. 2016.
      *Maria Sibylla Merian. Metamorphosis insectorum Surinamensium = Verandering der
      Surinaamsche insecten = Transformation of the Surinamese insects: 1705*. Tiel, Belgium/The
      Hague, Netherlands: Lannoo.
20   Merian went into the Labadist community at Waltha Castle with her mother and daugh-
      ters, where she stayed until moving to Amsterdam in 1691. For analysis of this period of
      her life, see Lölhöffel, Margot. 2016. Maria Sibylla Merianin und Johann Andreas Graff.
      Gemeinsames und Trennendes, Zweiter Teil. *Nürnberger Altstadtberichte* 41: 80–6.
21   Merian, 1679. *Raupen*, Chapter 1.
22   Appendix, study journal entry 2. See also p. 154 and p. 329 in this volume.

Marriage and the Move to Nuremberg

Maria Sibylla's marriage in 1665 to fellow artist Johann Andreas Graff of Nuremberg (1636–1701) was in many ways a providential match, although not a surprising one. She had known him from childhood; as an apprentice of her stepfather from 1653 to 1658 he would have been a familiar figure and potential mentor to her in her early artistic endeavors. Following his apprenticeship, Graff traveled and studied in Italy, and soon after his return to Frankfurt, he and Maria Sibylla were married. The wedding was a prestigious event; at least two booklets of poems were published to celebrate the occasion, and in one, friends and family of the couple contributed poems. In the 13-page booklet, a poem by Johann Andreas Graff's younger brother speaks of Maria Sibylla as an equal to his brother, at least in talent. Gottlieb Graff wrote:

> ... Here two [persons] make a loving couple, having come by painters'
>   favor
> and effort to [practice] equally the painters' art.
> How commendably 'twas done, how well it came to pass.
> Who could imagine such a pair as this united,
> as noble art, on the pathway of love –
> each the other's equal – has brought together! ...[23]

At the time of their marriage, Maria Sibylla was 18 and Graff was 29, and thus the equality presumed by Graff's brother speaks highly of her status as a young artist. A wedding poem by another contributor cautions the couple to avoid trying to "outshine each other in daubing."[24] Much ink has been spilled in speculation about the history and nature of the relationship of Maria Sibylla and her husband. The marriage ended in separation roughly two decades after this festive event, the birth of two daughters, and the publication of her first books. Graff eventually divorced her and remarried. Although it is unclear what roles Graff may have played in the production of her first books, it is certain that their move to his family home in Nuremberg shortly after the birth of their first child was a turning point for her work.

In 1668, Johann Andreas moved his young family into the Graff family house, where he had grown up and where his mother had lived until her death in 1649 (Figure 33).[25] When they arrived in Nuremberg, their first child, Johanna

---

23 Lölhöffel, 2015. Maria Sibylla Merianin und Johann Andreas Graff, Erster Teil: 48.
24 Ibid.: 49.
25 Ibid.: 51.

Helena, was less than a year old; their second daughter, Dorothea Maria, would be born ten years later.[26] The Graff residence was a former inn in the 'Milk Market' area of the city and was a spacious dwelling for the period. The house still stands, unusual in having survived the heavy bombing of Nuremberg in World War II. Three stories rise above the ground floor, and windows across the front of the upper floors provide a good deal of natural light, which would have been ideal for raising and painting insects. The ground floor likely served as a workshop, and it is large enough to have housed a press for printmaking.

The fourteen years spent in Nuremberg were among the most productive of Merian's life, even as she was raising two young daughters and maintaining this substantial house. In her time living there, she taught art to young women, produced three sets of flower plates and researched, wrote, and illustrated her first scientific work, the 1679 *Raupen* book. It is worth noting she would have been pregnant with her second daughter in the year before her first *Raupen* book was published; Dorothea Maria was christened February 3, 1678.[27] In Nuremberg, Maria Sibylla also completed most of the work of raising the insects that would appear in the 1683 *Raupen* book, and she studied several species that would be featured in her final caterpillar book, published decades later in 1717.[28] In total, she raised more than 100 species of insects through metamorphosis while living in Nuremberg. The city boasted an abundance of gardens,[29] which along with the surrounding meadows and forests, provided food and habitat for a wide variety of insect species.

In addition to providing ready access to her study subjects, the cultural aspects of the city helped to further Maria Sibylla's development as an artist/naturalist and provided abundant resources and a network of support that benefited her work. Nuremberg artists such as those of the Sandrart family were part of her network.[30] The environment of Nuremberg had produced a long list of renowned artists and artisans, including the painter Albrecht Dürer (1472–1528) and silversmith, Wenzel Jamnitzer (1508–1585). Less well-known but possibly

26   For more on Merian's daughters, who grew up to become skilled artists in their own right, see Reitsma, Ella and Sandrine A. Ulenberg. 2008. *Maria Sibylla Merian and Daughters: Women of Art and Science*. Amsterdam: Rembrandt House; Los Angeles: J. Paul Getty Museum.

27   Lölhöffel, 2016. Maria Sibylla Merianin und Johann Andreas Graff: 110, endnote 9.

28   Merian, Maria Sibylla. 1717. *Der Rupsen Begin, Voedzel en wonderbaare Verandering: waar in de Oorspronk, Spys en Gestaltverwisseling: als ook de Tyd, Plaats en Eigenschappen der Rupsen, Wormen, Kapellen, Uiltjes, Vliegen, en andere diergelyke Bloedelooze Beesjes vertoond word*. Volumes 1-3. Amsterdam: Published by the author.

29   There were more than 400 gardens in and around Nuremberg in Merian's time. Lölhöffel. 2015. Maria Sibylla Merianin und Johann Andreas Graff, Erster Teil: 60.

30   Painter Joachim von Sandrart (1606–1688) and his nephew Jacob von Sandrart (1630–1708), an engraver.

FIGURE 33
The Graff family house in Oberer
Milchmarkt (Upper Milkmarket),
Nuremberg

more influential on Merian was the painter Hans Hoffmann (1545–1591), another native of Nuremberg. He was clearly inspired by Dürer and made similar small naturalistic studies of plants and animals; however, Hoffmann also
created larger landscape paintings in which the animals and plants were the
main players and not merely accessories to a human narrative. His painting of
a hare resting in a patch of flowering plants is very different from other collages
of plants and animals produced at the time in the form of still life paintings
(Figure 34).[31]

Like Frankfurt, Nuremberg had a well-established publishing history, and
many botanical and other natural history books were printed in the city, home
to a number of naturalist/authors such as Joachim Camerarius the Younger
(1534–1598) and Basilius Besler (1561–1629). Paul Fürst was a Nuremberg publisher who produced books that would have been of particular interest to
Maria Sibylla: an edition of Hoefnagel's *Archetypa* and books of embroidery
patterns by his daughter Rosina Helena (1642–1709).[32] The pattern books

---

31   At least two more paintings by Hofmann of hares in naturalistic settings are known.
      Christie's London. 2015. *Old Master and British Paintings*. London: Christie's. Auction cata-
      log. December 8, 2015: 70.

32   See note 17, Chapter 1 regarding Hoefnagel's *Archetypa*. Hoefnagel, Joris, Jacob
      Hoefnagel, Paul Fürst, and Albert Pfähler. 1592. *Archetypa Studiaque Patris Georgii
      Hoefnagelii*. Nuremberg: Paul Fürst and Fürst, Rosina Helena. 1650. *Neues Modelbuch von*

FIGURE 34   Hans Hoffmann, 1582. A Hare among Plants. Watercolor and body color on
              vellum. A dandelion flower (lower right), plantain, garden tiger moth (upper
              right) and swallowtail caterpillar (centered above the hare) appear in Merian's
              first *Raupen* book.

were emulated by Merian to an extent in her earliest publications, as were aspects of Hoefnagel's compositions with insects. The literary life of Nuremberg
had an outlet in a group with indirect ties to her work and possibly a bearing on her decision to publish the *Raupen* books in German rather than Latin.
The *Pegnesischer Blumenorden* was a group of men and women who met to
read literary texts and further the culture of the German language. Christoph
Arnold (1627–1685) was an early member of this learned society,[33] and one of

---

       unterschiedlicher Art der Blumen. Nürnberg: Paul Fürst. Two other volumes followed this
       first one.
33   Lölhöffel, 2015. Maria Sibylla Merianin und Johann Andreas Graff, Erster Teil: 40–1.

Merian's most beneficial Nuremberg connections, contributing two poems to her first *Raupen* book.[34] Arnold was a scholar and book collector whose extensive private library included many natural history books.[35] Based on her references to some of these works in her own writing, is seems clear that she had access to these valuable resources. The ties between Christoph Arnold, his son Andreas, and Maria Sibylla are further documented by her contributions to the *album amicorum* (friendship album) of each man.[36]

Because 17th century Nuremberg was home to a broad community of people engaged in the arts and in natural history study, her simultaneous projects as naturalist, artist, teacher, and book publisher were supported in a variety of ways. An extensive network of family, friends, patrons, and professional contacts like Christoph Arnold provided her access to students for teaching, a ready audience for her books, and in some cases, material support for her work. Publicity and marketing opportunities were available through family connections such as Joachim von Sandrart, whose volumes of artists' biographies included Maria Sibylla and Johann Andreas Graff among numerous other entries. Recognition for the work of women was not uncommon in 17th century Germany; von Sandrart included a longer entry on Maria Sibylla than he did for her husband in his 1675 *Teutsche Academie*, and he included several other women in his accounts of Nuremberg artists and artisans.[37] Joachim von Sandrart's great niece, Susanna Maria von Sandrart, trained in her family's workshop and ultimately produced drawings and copperplate etchings/engravings.[38] The Nuremberg environment would have been encouraging to Merian. Women in the city regularly engaged in paid work across a variety

---

34   Merian, 1679. *Raupen*. Translations of these poems appear on pp. 134–36 in this volume.

35   Arnold's library is cataloged in Jürgensen, Renate. 2002. *Bibliotheca norica: Patrizier und Gelehrtenbibliotheken in Nürnberg zwischen Mittelalter und Aufklärung*, Vol. 1. Wiesbaden: Harrassowitz. Pp. 779–98 list his books on nature and medicine. Arnold used his extensive knowledge of natural history to compose laudatory poems included in Merian's first two insect books. See also Blom, Franciscus Joannes Maria. 1981. *Christoph & Andreas Arnold and England. The Travels and Book Collections of Two Seventeenth-Century Nurembergers. Ph.D. thesis, Catholic University Nijmegen.*

36   Maria Sibylla painted similar domesticated roses for both Christoph Arnold and his son, and these were entered into the *album amicorum* of each. The rose painted for Christoph Arnold included an inscription: "Man's life is like a flower." Lölhöffel. 2016, Maria Sibylla Merianin und Johann Andreas Graff, Zweiter Teil: 66–8.

37   Sandrart, Joachim. 1675/1679/1680. *Teutsche Academie der Bau-, Bild- und Mahlerey-Künste*, Nürnberg. The entries for Johann Andreas and Maria Sibylla Graff are on p. 339 of the 1675 volume. Sandrart knew Maria Sibylla's father from his time in Frankfurt, where he trained as an artist. According to the title page of Sandrart's book, Matthaeus the Younger sold copies of the *Teutsche Academie* in Frankfurt, further documentation of this connection.

38   Ibid.: 11.

FIGURE 35   Esther Barbara von Sandrart pictured with some
                   of her collection of naturalia. Engraved by Georg
                   Daniel Heumann in 1727 after a painting by Georges
                   Desmarées (1725)

of trades,[39] and a number of women in Merian's circle were respected artisans.
Esther Barbara von Sandrart (1651–1733), the wife of Joachim von Sandrart,
worked cutting jewels and cameos and was well respected as a collector of art
and naturalia. Below her engraved portrait is printed a tribute that honored
her even above her famous husband (Figure 35):

---

39   Leßmann, Sabina. 1993. Susanna Maria Von Sandrart: women artists in 17th-century
      Nürnberg. *Woman's Art Journal* 14 (1): 10. In many cases, the wife continued to run the
      family business after her husband's death.

Whose tireless work is applauded by Nature,
Whose noble oeuvre serves art of every kind,
For whose immense breadth of study one world is not enough:
her likeness is captured in this small picture.
But it does not capture her complete:
More illustrious than her famous husband,
this Sandrart is depicted by a genius all her own.[40]

Although not a member of the *Blumenorden* like Christoph Arnold, Maria Sibylla established her own informal circle, which she termed her *Jungfern Companie*, or "company of maidens." The group consisted of her painting students, daughters of both artisan and patrician families, who were not much younger than she was. Magdalena Fürst showed talent in drawing and painting, and after training with Maria Sibylla and Johann Thomas Fischer, she went on to color plates of important books and to paint watercolors of her own.[41] Clara Regina Imhoff and Dorothea Maria Auer were pupils and part of her 'company'[42] who also went on to become successful as painters. Maria Sibylla maintained close ties with some of her students for many years. The Graffs named their younger daughter for Dorothea Maria Auer, and Maria Sibylla continued to correspond with her after leaving Nuremberg. Letters written to Imhoff point to a special fondness for this patrician student. In 1697, she wrote to Clara Regina from Amsterdam and sent her a gift of the expensive pigment carmine, indicating that her former student was still actively painting.[43]

The younger women provided company for Maria Sibylla and, as art students, some income for the growing Graff family, but their family connections may have been even more valuable to their teacher in the long term. Through Clara Regina's great-uncle, who was principle guardian of the Nuremberg castle, Maria Sibylla was granted permission to use a garden near the castle, only two streets away from her home.[44] Another celebrated garden just outside the

---

40   The tribute was written by Nuremberg theologian and poet Joachim Negelein (1675–1749).

41   Doppelmayr, Johann Gabriel. 1730. *Historische Nachricht von den Nürnbergischen Mathematicis und Künstlern.* Nuremberg, 270–1. According to Doppelmayr, Fürst colored the plates of a great catalogue of plants in the gardens at Eichstätt as well as work by her father. Magdalena was the younger sister of Rosina Helena Fürst.

42   In a letter written July 25, 1682 to Clara Imhoff in Nuremberg, Merian refers to her "current company of maidens," indicating that she was again teaching painting after her move back to Frankfurt. Germanisches Nationalmuseum Nürnberg, Imhoff archive 1682-07-25.

43   Letter to Imhoff dated August 29, 1697. Stadtbibliothek Nürnberg, Autographen Nr. 167.

44   Appendix, study journal entry 119 a. See also Lölhöffel, 2015 Maria Sibylla Merianin und Johann Andreas Graff, Erster Teil: 62.

city walls belonged to Johann Christoph Volckamer, who was married to Regina Catharina Auer, sister of Maria Sibylla's pupil. Such gardens were a source of many of the insects and plants that she came to study, and she occasionally refers to these places in her study journal and in her published work, although not by the owners' names. Another branch of the Imhoff family had a noteworthy art collection, including works by Albrecht Dürer and Hans Hoffmann. It is conceivable that Merian's compositions incorporating plants and animals in a naturalistic setting were influenced by seeing works like Hans Hoffmann's hare, which is nestled among insect-damaged meadow plants and in the company of carefully portrayed insects and other small animals (Figure 34).

The Flower Books

In the fertile setting of Nuremberg, Maria Sibylla produced her first work for public consumption, a set of twelve etched plates of flowers (her first *Blumen* book), which the Graffs published in 1675.[45] Having been raised in the environment of a publishing house, she well understood the audience for prints of flowers in Nuremberg, a city rich with gardens. The garden culture spilled over into the needlework of artful women, hungry for new patterns such as those supplied by the books of Rosina Helena Fürst (Figure 36). Although her pattern books may have served as inspiration for Maria Sibylla's, the latter's plates were much more elaborate and naturalistic than Fürst's simplified designs.[46] Merian added elements of still life paintings to her *Blumen* book images, and she included insects in several of the plates, adding interest and liveliness to her compositions. The prints she designed and etched could be bound into a slender flower book, or they could be used individually as patterns for painting or embroidery.

The first set of flower plates, *Florum Fasciculus Primus*, was followed by the publication of another dozen compositions printed in 1677 and a final set of twelve in 1680.[47] Although she provided her married name below the title of each volume, this prominent space also was used to establish the familial connection to her father, Mattheus Merian. Johann Andreas Graff's name was included on the title page (Figure 37), another aid to sales, as he would have been known in Nuremberg for his cityscapes, landscapes and intricate

---

45   Merian, Maria Sibylla. 1675. *Florum Fasciculus Primus*. Nuremberg: Johann Andreas Graff.
46   Fürst, Rosina Helena. 1650. *Neues Modelbuch*. Nuremberg: Fürst.
47   Merian, Maria Sibylla. 1675–1680. *Florum Fasciculus Primus* [*Alter–Tertius*]. Nuremberg: Johann Andreas Graff.

FIGURE 36   Rosina Helena Fürst, 1650. *Model-Buchs Erster Theil.* This
                   title page features young women engaged in the socially
                   acceptable activity of needlework.

FIGURE 37   The left-hand title page is from Merian's 1675 (Part 1) *Blumen* book. Shelfmark Bip.
                   Bot.q.63#1, Staatsbibliothek Bamberg. The title page on the right is from her 1680 (Part 3)
                   Blumen book. Both state that Maria Sibylla Graffin is the daughter of Matthaeus Merian
                   the Elder, and that the book is available from Johann Andreas Graff. The wreath in the 1675
                   title page was adorned with insects whereas they are absent from the latter title page.

A. St Egydij Kirche zu Nurnberg B das Gijmnasium C Im Hoffische Häuser D Pellerische Häuser E der alte Eckelhof F der Plaß sonst der Dillinghof genant.

FIGURE 38   Johann Andreas Graff, 1682. Etching of Egidienplatz, Nuremberg

building interiors (Figures 38 and 39).[48] The flower prints must have sold well, because in 1680 the Graffs had the earlier plates reissued for sale to create a complete set of 36 prints (the New Flower Book or *Neues Blumenbuch*). In what was probably a calculated business decision, the flowers she depicted in her plates were not for the most part native to Germany, but were more exotic species, brought into the gardens of Europe from Asia (38 of 57 plants) or the Americas (5 plants). The tulip, brought from Persia circa 1554, was the most celebrated example of this botanical migration, but many more cultivars such as hyacinths, irises, anemones were to follow. The passionflower pictured in one plate was a later arrival from South America, entering Europe around 1610.[49]

---

48   Graff made his own copper plates in the early years in Nuremberg, but by 1685 employed his friend Johann Ulrich Kraus in Ausburg to etch/engrave the plates for the complex images that he drafted. Kraus had a family connection to Graff, being married to Johanna Sibylla, the daughter of Maria Sibylla's half-sister. Lölhöffel, 2016. Merianin und Graff, Zweiter Teil: 81.

49   Merian, Maria Sibylla, Thomas Bürger and Marina Heilmeyer. 1999. *Neues Blumenbuch* = New Book of Flowers. Munich: Prestel: 64–7. This facsimile volume of all three of Merian's flower books has helpful commentary and identification of all of the flowers in each plate.

Perfpectivifche Vorstellung def: Baugeriftes, wie solches im Jahr 1681. im anfang Iunij hej wieder auffrichtung der: Anno
1671. den 1 Octobris, bif: auff den Chor abgebranten Barfusser: Kirchen zu Nürnberg inwendig zu sehen gewest, und der.
gestalt abge Zaichnet und ins Kupffer gebracht worden, Von Johann Andreas Graffen Mahlern, bej deme es auch zu finden

FIGURE 39   Johann Andreas Graff, 1681. Etching of the reconstruction of the Franciscan
                 church in Nuremberg (Barfüßerkirche)

As the titles indicate, flower images were the focus of Maria Sibylla's earliest publications, and no text was included with the first 24 plates other than on the title pages. In the 1680 edition, she added a preface in German and an index of flowers listed by their local names. The use of German text could be attributed to the audience for the prints, young women who might not read Latin. The content of the rather elaborate 1680 preface is an interesting addition, and in it we see the possible influence of someone like Christoph Arnold. It begins with the tale of an old farmer planting date trees for future generations, and then takes to task those who spent and lost fortunes on flowers during the tulip boom that lasted from 1633 to 1637. The preface refers to other writers, including Bohuslav Balbín, and incorporates his description of an enormous Angelica plant in Bohemia. Part of the travelogue by Balbín was published by 1679 in Latin, and Merian may have known the work, but it is possible that Arnold, who read Latin, contributed this account and other more historical parts of the preface.[50] The 1680 preface then goes on to extol the joys of pleasure gardens, with particular praise for the one in Heidelberg pictured in an engraving by her father (Figure 25). The final paragraph of the preface of the 1680 *Blumen* book more closely resembles the voice of Merian in her other writings and is similar to the kind of information she included in the front matter of her scientific books. Here she states the intentions behind publication of the flower plates and inserts a bit of cross marketing for her 1679 *Raupen* book:

> But since now, in this springtime of abundant bloom and flower, Art is challenged as it were by Nature, as if to a spontaneous and graceful combat; it was both my desire and my obligation to bring some enjoyment to that contest, according to my abilities, willing though modest as they may be; and thus to publish this New Flower Book, not for my own use, as is done by others, but rather for the benefit of youth eager to learn, and also for the notice of posterity: so that it might be of service both for copying and for painting, for ladies' needlework, and for the use and enjoyment of all knowledgeable art lovers. In the confident trust that it will please these persons to receive such a three-volume Flower Book with that same gracious favor with which they received the recently published Caterpillar Book, for the flowers and plants to be found therein.

Perhaps most interesting is the poem that closes the preface, with further musings on the duel between art and nature:

---

50   Balbín, Bohuslav. 1679. *Miscellanea Historica Regni Bohemiae*. Pragæ: Typis Georgii Czernoch.

Thus must Art and Nature ever struggle with each other
until both sides shall gain such self-control
that victory is traded blow for equal blow:
the one defeated is no less the victor!

Thus must Art and Nature meet with fond embrace
and each one to the other extend her hand in peace:
Who battles thus does well! For after such a duel,
when all is finally done, both are satisfied.[51]

The "fond embrace" of art and nature is an apt metaphor for Maria Sibylla's life in Nuremberg. The preface to her first *Raupen* book, published a year before the 1680 *Blumen book*, acknowledged the dual nature of her readers, whom she addressed as "nature- and art-loving."[52] At the same time that she was producing the *Blumen* books, she was raising insects and recording their metamorphoses, research material that provided the basis for her first scientific book. Insects play a minor role in the *Blumen* books' plates, but a handful of images provide clues to the timing of her studies of metamorphosis. Like her stepfather, Maria Sibylla decorated several of her floral compositions with insects, and her first *Blumen* book features them on each of the 12 plates. Most of these are moths or butterflies, but there are two damselflies and a dragonfly, a mayfly or crane fly, a housefly and a beetle. On one plant, a spider dangles from its web, and this and other poses of the small animals are reminiscent of Marrel's compositions (Figure 40). Curiously, only nine more insects appear in the subsequent sets of flower prints published in 1677 and 1680, and five of these occupy a single plate featuring a bouquet of flowers in an ornate vase.[53] Not only are there many fewer plates with insects in the second and third sets, most were less carefully rendered than those in the first 12 plates. It appears that by the publication of the 1677 flower book, she was saving her lifelike images of lepidopterans for her 1679 *Raupen* book, in which the caterpillars and their life cycles would be the focus. Because the *Blumen* books were meant as model or pattern books, the simplified insect forms were preferable to the more detailed and complex images in her concurrently produced scientific work.

The plants and insects pictured in Merian's flower books are a mixture of those she observed directly and others that were copied from a variety of sources. Potential models for some of her insect images include paintings by

51   Merian, 1680. *Neues Blumenbuch*: preface.
52   Merian, 1679. *Raupen*: preface.
53   Merian, 1680. *Neues Blumenbuch*, Plate 3.

FIGURE 40   Plates 10 (violet) and 8 (iris) from Merian's 1675 *Blumen* book. Both plates feature insect
             damage to a leaf. In Plate 10, there is also damage to a flower petal on the upper right.

her stepfather and the aforementioned published images by Hoefnagel or
Jonston. Numerous flower books would have been available for copying, including the florilegium published by her father. Using the work of others as
models for illustrations was a common practice at the time and one used by
the Merian family firm and many other authors and publishers. One undisput-
ed source of material for the first *Blumen* book was the florilegium by Nicolas
Robert (French, 1614–1685), first published in 1660 (Figure 41). Her wreath on
the title page is copied largely from Robert's title page. Other parts of her compositions were only slightly modified from plates by Robert, who incorporated
adult butterflies into many of his graceful compositions, and in one case, a
caterpillar feeding on a leaf.[54] The iris appearing in Plate 8 of Merian's 1675 volume was modeled after one by Robert; however, her very naturalistic portrayal
of the swallowtail (*Iphiclides podalirius*) is not a copy from any source that I
have found, and it was likely painted from a specimen that she had on hand
(Figure 40). Her direct observations of plants and their flowers begin to make

54   Robert, Nicolas. 1660. *Variae ac multiformes florum species appressae ad vivum et aeneis
      tabulis incisae.* Paris: F. Poilly.

FIGURE 41   Title page (left) and Plate 14 (right) from Robert, 1660. *Variae ac multiformes florum species.*
             The iris by Merian in Figure 40 b is a close copy of Robert's flower here. The wreath in the
             title page of her 1675 *Blumen* book (Figure 60) also is very similar in composition to that on
             his title page.

an appearance in the *Blumen* book's plates, increasingly with the second and
third volumes of the set. Like Marrel and Robert, she chose to include insect
damage to several of the flowers in her plates, particularly in the first volume
(see Figures 40 and 42). The rose shown in Plate 19 of her caterpillar book
(p. 226) exhibits the same naturalistic pattern of herbivory as that seen in Plate
11 in her 1675 flower book (Figure 42), linking the images in the various book
projects that she completed while living in Nuremberg.

Beginnings of the *Raupen* Books

Plate 4 of the first *Blumen* book provides further visual evidence that by 1675
Maria Sibylla was raising caterpillars through metamorphosis (Figure 43). Most
of the narcissus plant was copied from the florilegium by Nicolas Robert,[55] as

---

55   Robert, 1660. *Variae ac multiformes florum*, Plate 23.

FIGURE 42
Plate 11 from Merian's 1675
Blumen book

was the adult comma butterfly, but she added an original image of the caterpillar of this species. She knew what the caterpillar of the comma butterfly
looked like, because she had been raising them, and proof of this is provided
by a matching image of the distinctively curved pose in entry 26 of her study
journal (Figure 70, p. 209). She reused this caterpillar image from the *Blumen*
book as part of the life cycle of the comma butterfly in her 1679 *Raupen* book
(Plate 14, p. 206), although she changed the appearance of the adult.[56] In her
Raupen book, she altered the flying form, and she added a second adult sitting
with wings folded, indicating her first-hand observations of this butterfly. In
the *Blumen* books she makes no effort to pair insects with their natural food
sources, but in Plate 14 of the *Raupen* book the caterpillar is shown feeding on
red currant, a plant that it will eat in nature. The *Blumen* book images were
primarily decorative and composed as such, whereas the *Raupen* book images

---

56   Merian, 1679. *Raupen*, Plate 14.

FIGURE 43   Narcissus and *Polygonia c-album* (comma butterfly). Robert, 1660. *Variae ac multiformes*
               florum, Plate 23 (left). Oak Spring Garden Foundation, Upperville, Virginia. Merian, 1675.
               Blumen book, Plate 4 (right)

were designed to show the life cycle of the insects and their relationship with
the caterpillars' food plants.

Three pieces of writing further verify that Merian was actively raising caterpillars even before 1675. In that year, Joachim von Sandrart described her
paintings of flowers, birds, and various insects and spiders and wrote that she:

> .... showed masterfully, with great deliberation, delicacy, and intelligence,
> in drawings, watercolors, and engravings, how they start out and subse-
> quently become living creatures, together with the plants from which
> they have their nourishment.

Sandrart closes the entry in his 1675 book by writing that she offered "her talents to the goddess Minerva in this way at the same time as managing her
household with great efficiency."[57] This glowing praise for Maria Sibylla's skills
and talents was published when she was only 28 years old. At least three years

---

57   Sandrart, 1675. *Teutsche Academie*: 339.

earlier, she had established such a reputation for collecting insects that in 1672, someone in Regensburg sent her a caterpillar as a gift.[58] If her study journal date of 1660 is reliable, she had experience with raising insects for 12 years at this point, but it is impossible to know the depth or breadth of her earliest studies. In her 1679 *Raupen* book she refers to a "five-year span of my investigations," indicating that she commenced her serious empirical studies around 1674.[59] However, it is important to understand that the work that comes to fruition in Merian's first scientific publication was built on her experience of two decades or more. Even as a young girl, she was learning how to portray both plants and insects with precision. Years of close observation of these subjects, in conjunction with the experience she gained by her successes and failures in raising them, informed the work she pursued for at least five decades.

Merian's Motivations

Various motivations for researching and writing about insects have been attributed to Merian, ranging from the financial to the purely pious. Many options for generating income would have been available to a talented artist like her, and spending countless hours collecting and raising insects and recording their habits in word and image would not be the most obvious career path for a young woman in the 17th century. That she did profit from her books and associated artwork is not insignificant.[60] This income made it possible to sustain her insect studies over time, but she did not enter into scientific pursuits for financial gain. Nor does the argument that her work was driven primarily by strong religious beliefs hold up under close examination. Her occasional use of pious language in her first caterpillar book and her time living with a religious community have led some to overstate the influence of religion on her body of work. One scholar postulated that the 1679 *Raupen* book showed that she was "interested in winning souls" by use of images that almost could be regarded

---

58   Merian's study journal entry 140 includes the following note: "A caterpillar of this kind (rather attractive in appearance) was sent in a little box from Regensburg to Nuremberg in 1672, as a welcome gift from the wife of the then Nuremberg emissary, and received by me; which (although I received it alive but did not understand or know at the time what its proper food was) died and spoiled, of no use to me."

59   Merian, 1679. *Raupen*, Chapter 33.

60   Ella Reitsma infers that Merian's combination of plants and insects was devised to sell books. Reitsma and Ulenberg. *Merian and Daughters*, 30. Others have addressed Merian's business acumen. See for example Kinukawa, Tomomi. 2011. Natural history as entrepreneurship: Maria Sibylla Merian's correspondence with J. G. Volkamer II and James Petiver. *Archives of Natural History*, 38 (2): 313–27.

as "devotional."[61] The timing of her relatively short sojourn with the Labadist community does not further the argument for pious motivations; she moved there after the publication of her second caterpillar book, and this may have been triggered by her family situation as much as by her beliefs.[62] The prevalence of natural theology in the 17th century makes it difficult to tease apart the study of nature from pious pursuits, but a comparison of Merian's writings with those of her near contemporaries is useful. Unlike Johannes Goedaert, she never compared the emergence of an adult butterfly from its pupa to the resurrection.[63] Goedaert related his insect studies to Christianity throughout his three volumes, and was exceeded perhaps only by Johannes Swammerdam and John Ray in linking his observations to the wondrous creations of God.[64]

Close examination of Merian's 1679 *Raupen* text shows that she touched upon the theme of natural theology much more lightly than many of her contemporaries. In the preface of her first caterpillar book, she makes her strongest statement of piety: "Therefore, seek herein not mine, but God's glory alone, and glorify Him as Creator of even these smallest and least of worms."[65] However, this passage could have been meant as insurance against criticism for presuming to elevate such lowly creatures; pious phrases like those of Merian's were still in use in the 19th century. She refers directly to God as a creative force only in the preface and in Chapter 32. In the latter text, however, she points to Nature as the mediator for God, and in four additional chapters, Nature is the creative force invoked. A striking passage in Chapter 32 echoes content of the image engraved by her father for Fludd (Figure 25). Here, Merian wrote that "God, through his handmaiden Nature, adorns many an inconspicuous and (as we might think) useless thing so wonderfully and beautifully."[66] In another chapter, she writes of the "diligent and artful love" of their offspring implanted into the creatures by Nature,[67] and further along in the book states that her composition of the colorful moth on the larkspur was designed "to delight the eye of appreciative admirers and to shed light on one small masterpiece of tireless Nature."[68] The inclusion in her book of a 'Caterpillar Hymn' by Christoph

---

61   Ludwig, Heidrun. 1998. The *Raupenbuch*. A popular natural history. In Merian and Wettengl. 1998. *Maria Sibylla Merian*: 62.

62   For a discussion of this period in Merian's life, see Davis, *Women on the Margins*, 157–66.

63   Goedaert, 1660. *Metamorphosis Naturalis*: 7–8.

64   See p. 32 in this volume.

65   Merian, 1679. *Raupen*: preface.

66   Ibid., Chapter 32. In this same passage, she writes that the colors on the moth are "artfully strewn on them, as if Nature had dusted it on."

67   Ibid., Chapter 22.

68   Ibid., Chapter 40.

Arnold adds several more references to God's creations, and this has been seen by some as further evidence of piety on Merian's part. In a second poem in the same volume, Arnold praises her accomplishments, and an alternative suggestion is that the Caterpillar Hymn with its religious tone was included to balance this exceptional tribute to a female naturalist.[69] Maria Sibylla undoubtedly held religious beliefs consistent with those of most German Protestants of her time, but her text does not indicate an unusually pious nature. Interestingly, by the time she wrote and published *Metamorphosis* in 1705, the mention of God in her writings had all but disappeared.[70]

Various motivations for her work are indicated in her 1679 preface. Merian wrote that after sharing her studies of insect transformations with people in Nuremberg, she was often "urged and encouraged by learned and respected persons – to present such Divine marvels to the world in a modest book."[71] This statement, thinly veiled with modesty and piety, provides a glimpse of her pride in having her work supported by people that she held in esteem. The motivations for publishing are different from those for conducting the research in the first place, as any scientist of the last five centuries would attest. For Merian, publication meant the possibility of income and something perhaps just as important to her, recognition for her work. In the closing of the same preface, she writes:

> The beginning has been made, and if it is your wish,
> I will pursue this work in service to my readers,
> that I may keep their interest keen with art
> and earn the praise and favor of great men.[72]

This statement speaks of ambition that is inconsistent with a "devotional" project. The acknowledgment of her readers and hints of future work and art to be produced show the same sort of business sense that comes through elsewhere in her writing, but she also reveals that she values receiving acclaim for her work. She wanted her work to benefit her readers, and in the 1680 *Blumen* book preface, she expressed her interest in "the notice of posterity."[73] Twenty-five years later in the preface of *Metamorphosis*, she proudly reports that she was "urged" to publish her work from Surinam by people who thought

---

69 See p. 138 in this book for more on Arnold's contributions.
70 The only mention of God in this later volume is in the conventional phrase: "if God grants me life and health...." Merian, 1705. *Metamorphosis*: preface.
71 Merian, 1679. *Raupen*, Preface.
72 Ibid.
73 Merian, 1680. *Neues Blumenbuch*, preface. See quotation on p. 62 of this volume.

it the "first and most remarkable work ever to have been painted in America." It is natural that, like many scientists before and since, Merian wished to share what she had learned. Even as a well-established naturalist in her fifties, she did not know if her Surinam volume would sell, but she was prepared to take the risk of the immense cost of publishing such a large illustrated volume. As she wrote, she was not seeking profit but "spared no expense to give pleasure and satisfaction to the art connoisseurs as well as to the amateurs of insects and plants."

For most humans, praise is uplifting and can provide additional incentive for work, but the years of effort behind her books led to the acclaim that she eventually received; however, wider acknowledgment of her work came later in her life, as described in Chapter 4 of this book. As a young woman, Maria Sibylla began collecting insects to paint, but the primary motivation for her lifelong study of insects is clear from her own writings: she was driven by the passionate curiosity characteristic of any scientist who pursues a field of research for years on end. She addresses her near obsession with insects in the preface of *Metamorphosis*, recounting her initial studies at an early age and then writing that:

> .... I also retreated from the company of people and engaged in this study. In order to practice the art of painting and to be able to draw and paint them from life, I accordingly painted all the insects that I could find, first in Frankfurt am Main and then in Nuremberg, meticulously on vellum ...[74]

It may be that the retreat into her studies contributed to the dissolution of her marriage. Certainly, the financially and physically risky journey to Surinam, and the challenges she experienced there speak to her single-mindedness in learning about new species of insects. Even the hardships of the tropics did not diminish her drive, and once she had published *Metamorphosis*, Merian turned her attention back to European insects, producing the third and final volume in that set.[75] In fact, in the decade after her return from Surinam, she was still making fresh observations, and entries in her study journal are dated as late as 1710, when she was 63 and possibly still infirm from a tropical malady. One example, reflecting her long interest in gall insects, is an entry in her study journal that was begun in 1684 and appended 26 years later: "This transformation is in my first *Raupen* volume, no. 50. In Amsterdam, in 1710, I again found this kind of oak medlar [gall], in which were such small black flies as the one shown here."[76]

---

74   Merian, 1705. *Metamorphosis*: Preface.
75   Merian, 1717. *Rupsen*.
76   Appendix, study journal entry 212.

If natural curiosity stimulated by her early studies kindled her passion for learning about insects, what sustained it through so many challenges? Again, Merian herself gives us the answer. Her glow of pleasure when she finally succeeded in obtaining all stages of a particular insect such as the Saturn moth is one of many examples: "Thus, when I did obtain it, I was filled with such great joy and was so pleased in my intent that I can hardly describe it."[77] As a biologist myself, I know this feeling well, and it is what drives many a scientific pursuit. In numerous places throughout her books, she expresses this sense of wonder and discovery, and it is clear that she found the long hours of work with insects rewarding on their own merits.

---

77   Merian, 1679. *Raupen*, Chapter 23.

CHAPTER 3

Described and Painted from Life

Searching fields and gardens for colorful caterpillars, carefully handling their soft, delicate forms, and providing them with armloads of plant food occupied Maria Sibylla Merian for at least 50 years of her life. Her homes were populated by wooden boxes in which the insects fed on plants, and changed from "worms" (*Raupen* or caterpillar) to butterflies or moths. Many, probably most, of her insects died before emerging from their pupae as adults, but ever tenacious, she was willing to spend years trying to raise at least one adult from each type of caterpillar. Each stage of metamorphosis was documented in word and image, as were the habits and behaviors of her small subjects. The food plants were also her subjects, and these as well as the insects needed to be collected, painted, and the plant names recorded. This diverse and copious raw material was compiled and edited into a published book, requiring finished compositions that combined insects with the plants on which their caterpillars usually fed. Likewise, the study journal notes were expanded to furnish the accompanying text. The copper plates used to print the images in her *Raupen* books were etched by Merian, and even the marketing of the books was part of her work. All of these steps were laborious, time-consuming, and were never guaranteed to produce a work that would sell well enough to repay the high cost of paper, copper, ink and the printer's bill. The beautifully illustrated book of insects and plants published in 1679 was undoubtedly the work of a talented and trained artist, but what made it distinctive in its time was its value as a scientific document. Examination of processes in each step from initial encounters with her subjects to the finished book will attest to the care that she invested in her research and the publication of her findings.

What we know about Merian's processes in making her *Raupen* book comes primarily from the text of the work itself, combined with some additional comments in her study journal and correspondence. In the preface to the first *Raupen* book she gives us a short history of her efforts, writing that while she sought to "adorn my flower paintings with caterpillars, summer-birds, and small creatures of that kind, as the landscape painters do with their pictures, to make the one, as it were, more lifelike by means of the other, I have also often taken the trouble to collect them …" Her intensive studies of insects and their plant hosts were the foundation of the work, and her training in painting and intaglio techniques established the skills and experience needed to produce the copper plates that illustrated her findings. Merian also read the works of other naturalists, and that background along with her own observations informed her text. In this chapter, we will follow her process from fieldwork to publication of her first scientific work in 1679.

Fieldwork: The Basis of Merian's Empirical Studies

Today, an entomologist or ecologist conducting research into a new species of insect has a wealth of resources, tools and technologies not available to the 17th century naturalist, but the overall process of investigation consists of steps similar to those employed by Merian.[1] Investigation of the life history of any species of plant or animal typically begins with finding and observing the organism in its environment. This important first step usually leads to more controlled observations, often in a laboratory setting. She began with her insect subjects in the gardens, orchards, meadows, and forests near where she lived, and her home served both as her laboratory and as her atelier. As a young girl, she most likely began by painting and then raising caterpillars found in the home garden or in that of a friend or family member. As she became more independent with age and the respectable cloak of marriage, Merian could forage further from home, and conduct what I will call her "field work." For example, in 1677 she discovered large numbers of caterpillars in the grass of an old moat in Altdorf, about 25 km east of Nuremberg. In the same study journal entry from that year, she mentions an area closer to home that she regularly searched for subjects, describing the small red beetles she found "... when I went up to my garden (next to the castle church, or Imperial Castle Chapel), both to see the flowers and to look for caterpillars."[2] She firmly asserts her dedication to these investigations in nature in the second paragraph of the 1679 Preface: "Not that it has not cost me

---

1   MacGregor, Arthur. 2018. *Naturalists in the Field. Collecting, Recording and Preserving the Natural World from the Fifteenth to the Twenty-First Century.* Emergence of Natural History, Volume 2. Leiden: Brill. This volume surveys the early history of field naturalists, their endeavors, and the challenges they encountered. As did Merian, naturalists from the Renaissance onward had to develop methods for collecting, preserving and transporting specimens, and as well as documenting their findings through notes, drawings and paintings, journals and published works. The chapters by Peter Barnard, Dominik Hünniger and Robert McCracken Peck specifically address investigations of insects.

2   Appendix, study journal entry 119. Her study journal image of the insect found in the dry moat shows only the larval stage. She reports that after feeding several of these caterpillars "daily from August into September," nothing came of them and they were "entirely spoiled." Thus, this insect was not included in her published work on European insects.

great effort and time to search for such small creatures and give them their food for many days, even months." For some species she exerted special industry in searching for the larvae that led to particularly striking adults, writing that she "spared no effort" to find the caterpillars of the large emperor moth and conducted "diligent investigation" to locate caterpillars of the small but brightly colored pease blossom moth.[3] Not all of her discoveries involved moths and butterflies; in 1679, she writes of her examination of the small red beetles that she noticed "on the red willows that grow along the water." Here Merian observed their eggs on the willow leaves "piled up together as if for a game of bowling," and reported that the warmth of the sun stimulated the larvae to chew their way out of the eggs.[4]

Maria Sibylla understood that beginning with caterpillars was the best way to investigate the life cycle of a moth or butterfly, and with few exceptions, she collected the larvae herself, at the same time noting on which type of plant she found them. The larval stages are much easier to see than the tiny and hidden eggs, and caterpillars are usually found on their food source, often with many on one plant. Caterpillars are slower moving, more abundant and thus much easier to collect in the field than adult moths and butterflies. She gives an example of this in Chapter 24 of her book, explaining that the adult yellow rose button moth "... is not easy to capture unless one obtains it by raising the caterpillar."[5]

Raising larvae through pupation to metamorphosis brings its own challenges. The first of these involves more fieldwork to collect the plants each species requires for nourishment, no easy task when growing larvae can eat double or triple their own body mass in one day. Combined with the fact that some larvae require one specific plant as their food, and others eat a limited range of host plants, gathering their food is a time-consuming occupation. While insects play the starring role in Merian's narrative, she did not neglect plants in her accounts of nature in and around Nuremberg.[6] Her commentary includes humble plants like ground ivy and dandelions, native trees and

---

3   Merian, 1679. *Raupen*, Chapter 13, *Saturnia pavonia* male (emperor moth) and Chapter 40, *Periphanes delphinii* (pease blossom moth).

4   Ibid., Chapter 27. *Chrysomela populi* (Pappelblattkäfer or red poplar leaf beetle) and *Salix caprea* (goat willow).

5   Ibid., Chapter 24. *Acleris bergmanniana* (yellow rose button moth).

6   Merian's interest in including indigenous plants in her book may have kindled by similar interests in other German naturalists, in particular those studying medicinal plants. Cooper, Alix. 2007. *Inventing the Indigenous. Local Knowledge and Natural History in Early Modern Europe.* Cambridge: Cambridge University Press: 36–8.

shrubs, and a number of domesticated varieties of fruit trees. It was essential to gather material from a wide range of these as well as from garden plants like roses and tulips, both to feed her hungry captives as well as to paint them to illustrate her work. She had a keen understanding of what would interest gardeners in her book, and she detailed the effects of insects on the various plants, such as the types of damage they inflicted on leaves, stems and buds. She wrote of extensive caterpillar damage to currant bushes in a local garden and devoted four of her chapters and plates to caterpillars that destroy the foliage and buds of roses. Merian remarks that this garden favorite, as well as many other flowers, often had more than one type of pest at a time.[7]

Her persistent hunt for both insects and their required food was essential to her ultimate goal of obtaining all stages of metamorphosis for her images, but in these days, months, and years exploring the pockets of nature around Frankfurt and Nuremberg, Maria Sibylla also made innumerable observations of her tiny subjects that touched on their behavior and ecology. For example, she noticed that the caterpillars of satin moths preferred to stay in the sun, whereas other species of caterpillars typically kept to the shaded side of leaves.[8] She documented the effects of weather and time of day on insects in the field, and behavioral patterns like the height and relative speed at which various adults flew. Many times her findings were the result of an intentional survey, such as her active search for eggs of the lackey moth.[9] She looked up in trees, scoured shrubs and herbaceous plants, but she also searched down low, and by doing so, she found caterpillars of the black cutworm in the soil. She also noted that black cutworms emerged at night to feed, indicating that her forays into nature were not limited to daylight hours.[10] Other than remarking on the difficulty in providing enough food for some caterpillars, she gives us no sign that she viewed her work as drudgery, but instead, expresses overt pleasure in many of her discoveries. The looping movements of the peppered moth caterpillar are such that "one could hardly watch without delight," and the similar movements of the caterpillar of the garden carpet moth are "almost comical."[11]

---

7   Ibid., Chapter 29. Merian describes the devastating damage on currant bushes. Roses are featured in Chapters 19, 22, 24 and 28.
8   Ibid., Chapter 30. *Leucoma salicis* (satin moth) on *Salix discolor* (pussy willow).
9   Ibid., Chapter 33. *Malacosoma neustria* (lackey moth) on *Prunus spinosa* (blackthorn or sloe).
10 Ibid., Chapter 43. *Agrotis ipsilon* (dark Sword-grass moth or black cutworm).
11 Ibid., Chapter 37. *Biston betularia* (peppered moth) and Chapter 12. *Xanthorhoe fluctuata* (garden carpet moth).

Laboratory Work

What we visualize as a biological laboratory today did not exist in Merian's time, and natural history investigations were usually conducted in domestic spaces and even gardens. It is appropriate to use the term laboratory to designate the place where she studied insects, just as one would refer to a home studio or atelier. As well as being the family living space, her home housed her artistic and scientific endeavors and the family business.[12] In all probability, multiple spaces in the large house in the "upper milk market" of Nuremberg were used for her work, possibly changing location with time of year, temperature and available light. The front of the house on *Oberer Milchmarkt* (presently Bergstrasse 10) faces south-southwest, and the windows on that side of the house provided daylight and ventilation, helpful for insect husbandry, observations and artwork. In the 1679 *Raupen* Merian mentions a dozen species of larvae or pupae that she kept over the winter and notes that several of these emerged as adult moths or butterflies early in the warmth of her house.

Often it is difficult to know whether she made a specific observation in the field or in her "laboratory" at home. She recorded differential feeding on various parts of plants, the caterpillars' movements, and their defensive responses to being touched. These and other behaviors that she observed could have been displayed both in captive and free larvae. Merian does not tell us in most instances where she witnessed various phenomena. On the other hand, the physical transformations of caterpillars to pupae and the eventual emergence of the adults must have occurred in most cases in her home laboratory, where she could have kept a close watch to avoid missing the changes in her subjects.

Perhaps the best way to understand Merian's process would be to imagine what her working day could have been like in the busy growing season of spring, when the largest variety of caterpillars would be feeding in and around Nuremberg (Figure 44). After taking care of her household duties, she might

---

12   Natural philosophers before and after Merian, including Charles Darwin, Johannes Swammerdam and Galileo conducted their studies from home, and this was the norm for centuries before science became a professional pursuit with dedicated space at a university or elsewhere. The term "home laboratory" has been used to describe the workspace of these men, and it is reasonable to apply this term to the spaces where Merian spent countless hours in the study of insects. For more on this topic see Opitz, Donald L., Staffan Bergwik, and Brigit Van Tiggelen. 2016. *Domesticity in the Making of Modern Science.* New York: Springer and Werrett, Simon. 2019. *Thrifty Science: Making the Most of Materials in the History of Experiment.* Chicago: University of Chicago Press. Darwin was a relative latecomer to the use of his garden as a laboratory, a practice documented from medieval times onward. See for example Baldassarri, Fabrizio. 2017. Introduction: gardens as laboratories. A history of botanical sciences. *Journal of Early Modern Studies*, 6(1): 9–19.

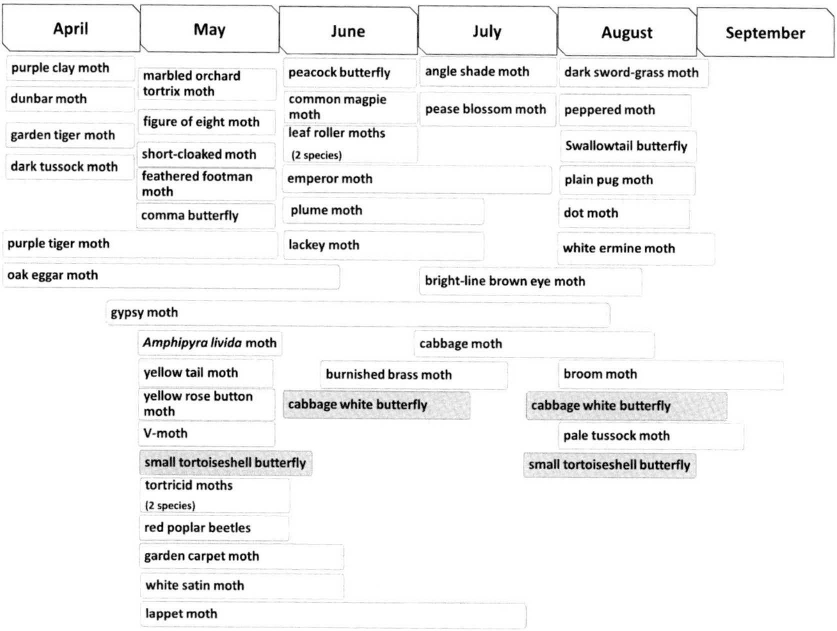

April	May	June	July	August	September
purple clay moth	marbled orchard tortrix moth	peacock butterfly	angle shade moth	dark sword-grass moth	
dunbar moth	figure of eight moth	common magpie moth	pease blossom moth	peppered moth	
garden tiger moth	short-cloaked moth	leaf roller moths (2 species)		Swallowtail butterfly	
dark tussock moth	feathered footman moth	emperor moth		plain pug moth	
	comma butterfly	plume moth		dot moth	
purple tiger moth				white ermine moth	
oak eggar moth			bright-line brown eye moth		
	gypsy moth				
	Amphipyra livida moth		cabbage moth		
	yellow tail moth	burnished brass moth		broom moth	
	yellow rose button moth	cabbage white butterfly		cabbage white butterfly	
	V-moth			pale tussock moth	
	small tortoiseshell butterfly			small tortoiseshell butterfly	
	tortricid moths (2 species)				
	red poplar beetles				
	garden carpet moth				
	white satin moth				
	lappet moth				
	lackey moth				

FIGURE 44   Chart indicating the timing of Merian's husbandry for each type of larva featured in her 1679 *Raupen* book. Grey bars indicate species that reproduced at two different times in one season. Timing is based on information in her chapters.

climb the hill and steps to the castle garden that she used as one source of insects.[13] Alternatively, she might need plant material not available in that garden, which would necessitate a further trek to a meadow or a larger garden with an orchard. Once at her field site, she would inspect likely plants for new insects and begin collecting them, perhaps in small cloth bags. On this same trip, she could begin to fill her basket with cuttings of plant food for insects already waiting hungrily at her house. On the walk back, she might gather more plant material, being careful to select the foods she knew would fit the preferences of her growing caterpillars. Upon returning home, she would set up any new species into separate wooden boxes. Merian succinctly describes this housing, writing, "I kept all my caterpillars, large or small, many or few, in a box, having first made a number of small holes in it so that they could have air."[14] Each species of caterpillar would have been maintained in a separate box in order to avoid any confusion about identity of the metamorphic stages. In May, she might be confronted on any given day with a dozen or more

---

13   See p. 58 in this volume for more on this garden.
14   Merian, 1679. *Raupen*, Chapter 30.

wooden boxes, each with a different type of caterpillar, and many of these small animals requiring very different food plants. Each box of insects would need to be supplied with fresh plant material every day or two, because as she was aware, the moisture in plants serves as the water source for most caterpillars. However, she also reported on three types of caterpillar prone to drinking water, so she had to supply that as well as food to assure their survival and successful metamorphosis.[15]

As Merian replenished the larval plant foods, she could observe and make notes on the kind of feeding patterns exhibited, and whether, when offered a new type of food, the caterpillars would accept it. At times, it must have been difficult to keep up with their incessant demands, because for two species she reported that they would cannibalize each other when she did not supply enough plant food.[16] In less dramatic circumstances, she tried substituting a different plant for the usual food. Some caterpillars were found to be more particular than others in their choice of plants; her study journal notes and published books describe a wide variety of such examples. In addition to refreshing the food of the insects, frequent cleanings of the boxes would have been needed to remove frass (insect feces), any dead insects, and dried out or spoiled plants. The waste produced by several feeding caterpillars can mount up quickly, and like rotting plant material, the frass, rich in organic material, is prone to support fungal growth, which can spread to the insects and kill them.

Merian would need to take care not to accidentally lose any of her small creatures in the cleaning process, as they cling tenaciously to plant material and can be camouflaged and easily overlooked (see for example Plate 36 on p. 294). Sometimes new and unexpected inhabitants would appear in her boxes. In speaking of the caterpillars of the peacock butterfly, she wrote, "I placed them in a large box, and one time when I forgot to clean it out I noticed on the bottom, underneath their droppings, white maggots or worms, some of which had already changed into black capsules."[17] In this instance, the flies were not harmful to the butterflies, but this was an unusual case. Most flies or wasps found in with the lepidopterans would have been deadly parasitoids, as we shall see; unbeknownst to Merian these invaders would have been deposited as eggs into the bodies of the caterpillars while they were still in nature.

In addition to the work required to maintain her insects in a healthy state, Merian would have used her laboratory time to make observations of their

---

15   Ibid., Chapter 21. *Lasiocampa quercus*, the oak eggar moth is one example of a type of caterpillar that Merian observed drinking water.
16   Ibid., Chapter 23. *Saturnia pavonia* female (emperor moth) and Chapter 35, *Melanchra persicariae* (dot moth).
17   Ibid., Chapter 26. *Aglais io* (peacock butterfly) and *Urtica dioica* (common nettle).

behaviors and metamorphoses, recording distinctive types of locomotion and interesting defensive postures whenever there was time. If any signs of metamorphosis had begun, she would need to record an image of this as soon as she could prepare her pigments, vellum and brushes. These dramatic transformations from caterpillar to pupa or the wondrous process of adult emergence were transitory, and documenting these changes needed to take priority if she were to capture them in time. As soon as possible, she would have written up her notes on what she observed, providing further material for her book. Her description of the four to five hours required for pupal formation in the peacock butterfly gives us an example of the timing notes that Merian must have made; she also painted the emergence of the adult peacock butterfly and recorded the fact that it took half an hour for the wings to expand and stiffen.[18] In many instances, she would have had to alter her schedule and put everything else on hold to capture an image, as on the day that she received a lappet moth caterpillar from some unspecified person. This large caterpillar began to enter pupation almost at once, requiring her to forgo sleep that night in order to observe and to paint the change.[19]

Once metamorphosis was complete and Maria Sibylla had a moth or butterfly in hand, she could afford to wait to paint the adult. When she needed a specimen as a painting model or wanted to preserve one in her collection, she killed it quickly with a hot needle and then could spread its wings and pin it in a box.[20] Most, but not all, of her adult moths and butterflies were depicted with spread wings, and for those with folded wings (e.g. Plates 39 and 44 in the 1679 *Raupen* book) she may have preserved them in a different position, or perhaps painted them when they were sitting still.

On some days, Merian would pause in her usual work to take a closer look at her subjects, perhaps with a magnifying glass or by cutting open a structure to examine the contents. She was very curious about what was occurring within the pupae, and in her first caterpillar book she describes in two places how she used dissection in her examinations.[21] Her invasive investigations did not stop

---

18   Ibid.
19   Ibid., Chapter 17. *Gastropacha quercifolia* (lappet moth).
20   Within a letter written in 1697 to her former student, Clara Regina (Imhoff) Scheurling, Maria Sibylla described how she killed the butterflies in this fashion, a method that was quick and preserved the wings intact. Autograph Number 167, Stadtbibliothek Nürnberg. Her collection might have included some unbroken pupa and cocoons, but probably not the caterpillars or eggs, as those would have required liquid preservative. Merian, 1679. *Raupen*: 2.
21   Merian, 1679. *Raupen*, Chapter 2 and Chapter 5.

with pupae, and Merian also cut open plant galls to satisfy her curiosity about what formed them and what could be found inside.[22]

If she was lucky in the timing of her observations, she would have seen adult females laying eggs. The probability of witnessing egg laying would have been increased for species of moths that mate and lay eggs very soon after the adults emerge. Even if she did not observe oviposition, she often noted the presence of insect eggs and the emergence of the tiny caterpillars as they chewed their way out, ready to begin the next stage of their life cycle. Again, these transitions were ephemeral phenomena and it would have been necessary for her to capture the details in notes and a painted study image right away in order to remember the details, especially during a period when she had many different species growing in her laboratory.

Many times, Merian's attempts to raise a species came to naught, because the insects died or were killed at a stage that left her with an incomplete metamorphosis. Undaunted, she would begin again the next year with the caterpillars as they appeared, and several species took her multiple years of effort to achieve an adult. Even when the process was interrupted and restarted later, she would ultimately combine her observations, sometimes made over years, into a seamless account. The complete life cycle of the dot moth (Chapter 35) and the cabbage moth (Chapter 49) each took her four years of trial and error, and she finally was successful with the emperor moth after three years of attempting to raise it through metamorphosis. Undoubtedly, other species took her more than one year of effort to obtain the adult, but she did not document this in every case.

To gain an appreciation for the challenges of successful insect husbandry, we may compare Merian's successes in raising dozens of species to the survival rate of insects raised in a modern research laboratory, in which a number of scientists were working with a few species. Professional scientists investigating two of the same species of butterfly studied by Merian achieved a survival rate for caterpillars of the small tortoiseshell butterfly and the peacock butterfly of 34% and 15%, respectively.[23] A common cause for failure in attempts to raise lepidopterans can be traced to attacks on caterpillars and pupae by parasitoids. If flies or wasps appeared in the wooden boxes, Merian documented their stages and appearance along with those of her primary subjects. In her 1679 *Raupen* book, she included parasitic flies and wasps in

---

22   Ibid., Chapter 42. In 1684, she dissected more galls, as described in her study journal
      entry 212.
23   Audusseau, Hélène, Gundula Kolb, and Niklas Janz. 2015. Plant fertilization interacts with
      life history: variation in stoichiometry and performance in nettle-feeding butterflies.
      PLOS ONE, 10(5): 9.

eight of the fifty chapters and plates, and in two cases, the butterflies were host to more than one species of parasitoid.[24] Other than the flies and wasps whose larvae made their living off the bodies of the lepidopterans, she also noticed and wrote about various independently living flies (not parasitic) of at least three types: the drone fly, gooseberry sawfly and some flies in the Phorid family. Two species of beetles and the various inhabitants of galls found on a black poplar round out the collection of insects portrayed in her first *Raupen* book.

Merian's Study Journal

Although we do not know at what precise point she began to take notes and make images in support of her research on insects, Merian documented 1660 as the beginning of her independent investigations. The bound manuscript that today resides in St. Petersburg represents all that we now have of her research notes, and these sheets are dedicated to her insect subjects, showing little plant material. This manuscript has alternately been referred to as her study book (*Studienbuch*), or her book of notes and studies, but I will use the more appropriate and concise term study journal here. The existing manuscript consists of 133 pages of paper with handwritten entries and 318 pieces of vellum with her studies. The vellum pieces are irregularly shaped and are most likely remnants trimmed from larger pieces. She painted on vellum using watercolor and body color, which is watercolor made opaque with the addition of lead white. Numbered text entries are on one side of a page and the facing page exhibits the attached studies, numbered by Merian to match the text. The studies are mounted up to three to a page in vertical format; the larger studies are mounted one or two to a page. For every insect depicted, the stages in these studies are painted life-size, so the number of insects on one piece of vellum varies with the dimensions of the scrap and the size of the animals.

The study journal entries pertaining to the 1679 *Raupen* book can be found within the first 50 sheets of the handwritten manuscript. Curiously, she put all of these early entries down on paper in a careful calligraphic hand well after the publication of her first two *Raupen* books in 1679 and 1683, and they clearly are not the original notes documenting her investigations. The prevailing school of thought is that she dedicated some of her time in residence at Waltha Castle with the Labadist community to ordering and recopying her original research notes. This conclusion is based upon the fact that almost all of the

---

24   Merian, 1679. *Raupen*. Parasitic flies and wasps appear in Plates 5, 22, 26, 28, 33, 44, 45 and 49.

entries in these pages refer to the chapter and book in which they already had been published (see sample entries in the Appendix of the present book).[25]

If her original investigations began in 1660 as she states on the initial page of this manuscript, nineteen years elapsed between this time and the publication of her first *Raupen* book, and at least three or more years went by before she began to recopy her early notes during her stay at Waltha Castle from 1686 to 1691. Merian's original notes probably were made as she worked with her insects at home or returned from the field, and were most likely quick and perhaps shorthand notations such as any busy scientist would jot down. We may never know her motivation for rewriting these entries in such a time-consuming and laborious calligraphic hand, but it has been proposed that this was done to organize her reference material, and perhaps to leave a tidy record for those who might follow her. I would add that this was perhaps the one period in her life during which she had the time to do this, as she was not as actively pursuing her empirical research while at Waltha as she had done prior to joining the community there. The remainder of the study journal existing today contains entries that were made after she left the Labadist community; many of these pertain to her Surinam volume and these will not be addressed here.

In her written study journal entries, Merian typically stated when she collected an insect, but these dates are frustratingly incomplete and usually consist of the month and sometimes the day, but rarely the year. The timing of metamorphic stages was often included as well, however she wrote for many species that they pupated for about two weeks, so this information is lacking in the specifics desired by a modern biologist. For most species, she reported the type of plant on which she found the caterpillars, and often, their feeding habits. In addition to this baseline information, she tended to insert observations that could not be portrayed in the accompanying image, such as behaviors, movements, egg location, and how the pupae were attached to a substrate. Merian also used the study journal to record effects of weather on some insects, where and when adults would fly, and information on the parasitoids (although she thought of them as "false transformations" at the time).

---

25   Merian and Beer, *Leningrader Studienbuch*: 39. All 133 leaves with texts and the mounted watercolors are on the same kind of paper, which measures 20 × 22 cm. The manuscript also contains fourteen pages with images but no text. The complex history and analysis of the study journal was constructed for the commentary of this facsimile volume by a team of experts who analyzed paper, handwriting and the organisms. My opinion is that it is unlikely that Merian would have had the time to execute the calligraphic version of the first third of the manuscript while she was actively conducting her research in Nuremberg. The further disruption of her move from Nuremberg back to Frankfurt, the disturbance in her marriage and her move into the Labadist colony combine to reinforce the idea that Beer's chronology makes sense.

FIGURE 45   Merian. Study journal painting 8o. *Cosmia trapezina* (dunbar moth),
              Xanthorhoe fluctuata (garden carpet moth) and two unidentified species.
              On the left is a caterpillar from which she never obtained a pupa or adult
              and below this is a parasitic insect. The middle series of the dunbar moth is
              depicted in Plate 3 of the 1679 *Raupen* book, and to the far right are the stages
              of the garden carpet moth found in Plate 12 of the same book.

The study journal entries in the Appendix are representative in content, and
these sample entries demonstrate the range of length and depth of her handwritten notes.

    Merian's painted studies of her insect subjects fell within certain self-
imposed constraints, parameters largely dictated by her desire to make her
insects readily identifiable to her readers. The accurate portrayal of the insect
stages relied upon study images that were painted upon immediate and close
observation of eggs, larvae, and pupa. Specimens of adult moths and butterflies last for years if properly pinned and protected from light and the ravages
of pests, but it is difficult to preserve the earlier stages, i.e. the caterpillars, in a
way that maintains their form and color. A sampling of representative paintings from her study journal may be seen in Figures 45, 46 and 47. She included
evidence of several failed metamorphoses in her study journal in the form of
orphaned caterpillars populating mostly empty pieces of vellum. Sometimes
she depicted a caterpillar and pupa but no adult (Figure 45). As a subsample of
these thwarted studies, there are fourteen images of incomplete lepidopteran
metamorphoses appearing in the first 90 entries of her journal. Of these, seven
of the caterpillars pictured are not mentioned at all in her study journal text,
but for the others she included both a small painting and some cursory notes.

For example, study journal entry 80 describes a caterpillar that was parasitized by the insect with which it shares the vellum, and she reports that the larva lived 14 days without eating.

We must keep in mind that each piece of vellum in Merian's study journal is like a time-lapse photo, in that the stages depicted would not have existed at the same time. For most life cycles, she would have begun by painting the caterpillar and would add the pupal stage later. If she were fortunate enough to have a successful metamorphosis, an adult would be painted; eggs were included if she observed them. When insects overwintered as caterpillars or pupa, her painted studies and notes were interrupted for several months. If she lost larvae or pupae to a fungal or parasitic infestation, she began again in another year with fresh caterpillars, and it appears that she saved space on the original bit of vellum to paint the hoped for adult moth or butterfly. A finished study journal painting could have been created over any length of time, from a stretch of a few weeks to a few months or even years, and the same is true for her written notes.

Because each set of written notes and images were appended over time, Merian used a numbering system to keep associated texts and images linked and her observations organized for use in her published work. Due to the staggered nature of the stages of metamorphosis, she would have needed to keep track of each set of notes and images in order to summarize the sequence of information for each species of insect correctly. To match up her notes, images and living stages of the insects, she must have labeled her wooden boxes, perhaps with the same numbering system that she used to link the text and study paintings. If this were the case and she used sequential numbers throughout the year, then one might expect the study journal entries to reflect a clear chronology, but it is not apparent that they do. The somewhat disjointed order of Merian's research notes can be explained in part by the fact that many of these entries were assembled over a period of years.

Any attempt to establish a timeline for entries in Merian's study journal is further complicated by the fact that she frequently combined more than one species of insect in a single numbered entry and in the associated painting on vellum. The number of specimens that she depicted in one small painting was largely dictated by the size of the vellum scrap, because she painted all insects as life-size. Thus, for many insects with smaller adults and larvae, two species share one piece of vellum. Study journal entry 24 is one such example, in which the feathered footman moth is shown on the left-hand portion of the vellum and on the right is the tabby moth (Figure 46). Her notes for entry 24 describe the feathered footman in the first paragraph, noting that it appears in her first

FIGURE 46   Merian. Study journal painting. *Spiris stirata* (feathered footman) eggs, pupa,
              cocoon and caterpillar shown on the left are the subject of Chapter 15 in the
              1679 *Raupen* book. On the right side of the same piece of vellum, Merian
              painted the stages of *Aglossa pinguinalis* (tabby moth).

Raupen volume in Plate 15. She then describes the tabby moth in a separate
paragraph, reporting that it appears in Plate 4 of her second *Raupen* volume.
Conversely, the spatial and temporal sequence for study journal entry 66 is
reversed (Figure 47). The dark arches moth pictured on the left was featured
in her 1683 *Raupen* book, and the small Tortricid moth on the right-hand portion of the vellum is pictured in Plate 7 of her 1679 *Raupen* book. Likewise, the
text describing the left-hand moth appears in her notes above the text for the
Tortricid moth, even though the latter is described in print four years earlier in
1679. This "reversed" sequence occurs elsewhere in Merian's study journal. In
such instances, she seems to have used the left side of the vellum to begin the
first painted study, and the second set of specimen images was added as space
allowed. Perhaps she did not obtain the complete metamorphosis of the dark
arches moth in time for inclusion in her first *Raupen* book, and she completed
the notes and study painting once she had all of the stages. Alternatively, she
may have chosen to leave the dark arches moth for her second *Raupen* book for

FIGURE 47   Merian. Study journal painting 66. The *Apamea monoglypha* (dark arches
               moth) on the left is featured in Plate 26 of her 1683 *Raupen* book, and the small
               Tortricid moth on the right is pictured in Plate 7 of her 1679 *Raupen* book.

other reasons. Once again, Merian leaves us with a mystery, and we may never
know precisely how she kept track of dozens of sets of notes of studies in various stages of completion over several years.

   An additional mystery is the fate of Merian's plant studies. Evidence of the
existence of such paintings as late as 1711 comes from a firsthand account by
Zacharias Conrad von Uffenbach. No one knows the eventual fate of these
plant studies, but when von Uffenbach visited her at her home in Amsterdam,
he saw a number of insect images and specimens and "a very large *Volumen*
thicker than your hand in which were all kinds of plants and fruits, both for-
eign and European, also painted from life, all on parchment."[26] In order to produce images integrating the insects and plants that ultimately appeared in her
finished plates, Merian would have needed source material. In all likelihood,
the plant studies were sold off after her death along with the insect studies,
notes, copper plates and the rights to the books.

---

26   Uffenbach, Zacharias Conrad von, 1753. *Herr Zacharias Conrad von Uffenbach Merckwürdige
     Reise durch Niedersachsen Holland und Engelland.* Ulm: auf Kosten Johann Friedrich
     Gaum: 553. He visited Merian on 23 February 1711.

Merian's Other Sources of Information

The contents of Merian's European caterpillar books consist almost entirely of her own first-hand observations. She makes this fact clear by writing in the first person and active voice to tell readers how she went about her work: "I have found ...", "I have had small caterpillars ..." and "I have often observed...." The significant exception to the use of first person voice is in the entry regarding the silkworms, and as discussed later in my commentary on Merian's text, it appears that she is presenting information given to her by breeders of this economically valuable, but non-native, species of moth. Nowhere in her 1679 *Raupen* book does Merian mention the name of another naturalist, although Christoph Arnold does so in the celebratory poem written for the book, referring to Goedaert and Swammerdam among others.[27] Merian's indicates in her text that she had read other works on insects, referring in one chapter to "what that scholar wrote" and in several instances including descriptions of moths or butterflies that are similar to those of Johannes Goedaert.[28]

Examination of Merian's images make it obvious that she did not copy material from Goedaert's plates (compare Figures 3 and 4 in Chapter 1). However, comparisons attest to the fact that she raised some of the same species that he did, and she occasionally observed similar behaviors. Her language conveys a less overtly religious tone than Goedaert's writings, and she contradicted some of his conclusions, in particular, those in which he invoked the spontaneous generation of insects. Merian improved a great deal on Goedaert's descriptions of the insects, adding information about colors, and unlike him, she included in her books only moths and butterflies for which she had larvae, pupae and adults.[29] The biggest difference from Goedaert in the images and overall content, however, is the emphasis Merian places on the larval host plants.

It is not known from where Merian derived the Latin plant names she applied in her book. However, an enormous variety of herbals and florilegia were in print by the first half of the 17th century and many of these included recognizable plant images labeled with the names that she used. For plants like *Atriplex* (Chapter 41 in 1679 *Raupen*), the Latin names she employs can be found in the much older writings of Hippocrates, Pliny, and Dioscorides.

---

27   See p. 134 in this volume for the translation of Arnold's poem.
28   Merian, 1679. *Raupen*, Chapter 37. The unnamed scholar referenced by her could have been Jan Jonston, whose work was published by Maria Sibylla's half-brothers. Jonston, Merian and Merian, *Historiae Naturalis de Insectis*.
29   Goedaert depicted the caterpillar, pupa and adult for 49 out of 102 lepidopterans featured in his three volumes. Merian's three books on European insects and her Surinam volume included the caterpillar, pupa and adult for at least 200 species of lepidopterans.

These plants appeared in herbals printed in Frankfurt, Nuremberg, and across Europe in the 16th and 17th century.[30] No definitive evidence of the source of the plant names that she used has been established, but she had access to botanical books through the family publishing house as well as through acquaintances such as Christoph Arnold, who had an extensive library of herbals and other books on plants.[31] Merian often supplied her readers with the local plant name in German as well as the more widely accepted Latin name. As mentioned in the first chapter of this book, no such analogous set of insect names, either common or Latin, existed for species when she published her first *Raupen* book, nor was this the case in entomological works for decades to come. In a few of her chapters she informs readers of local names for insects that she has learned from gardeners, such as *Beermutter* for the caterpillar of the garden tiger moth.

Illustrating the *Raupen* Book

We have Merian's writings to guide us in understanding how she conducted her investigations, recorded her findings and made early images of her research subjects, but more variables are introduced in reconstructing the steps that she took in order to illustrate her book. Art historians have discussed the aesthetic nature of her images, but here I wish to explore how the information content in the *Raupen* book's plates functioned hand in hand with the text. I also will discuss her process in creating the images. As with many factors regarding Merian's history, we have no definitive record of her working methods and so many conclusions drawn here are educated conjectures.

In representing plants and animals in an image, every artist makes choices and thus interprets nature in their own way. The first half of the title of the 1679 *Raupen* sets up Merian's central choices for us, informing her readers that "by means of an entirely new invention" she will portray "the wondrous transformation and particular food plants of caterpillars." As a skilled and talented artist, Merian was capable of rendering her images much as she conceived them, and by her accurate and detailed portrayal of the organisms, she conveyed a great deal about their structure. Her images were the earliest to go further than

---

30   For more on the example of *Atriplex* see Kusukawa, *Picturing the Book of Nature*: 112–14.

31   Jürgensen, *Bibliotheca norica*: 783–7. Arnold's collection encompassed numerous botanical books such as those by Mattioli (see p. 4 in this volume); such herbals are illustrated and have most of the Latin plant names used by Merian. Arnold's other natural history books included Moffett's 1634 book on insects and Goedaert's 1660 *Metamorphosis Naturalis*.

this however, showing for the first time the interactions between ecologically related organisms. Insects are no longer objects taken out of their natural context, but instead actively engage with their environment.

Merian's placement of the plant in the middle of most plates would seem to bring focus to the botanical offering, but in fact, the vegetation may be viewed as playing a supporting, albeit critical role, to the actions of the insects on and around it. By centering the plants on the page, she followed the conventions of her time, and this well-established compositional device makes it possible to fit more of the plant on a page. The primary subject of her book – the caterpillars and their transformations – are smaller and usually could be arranged around the central plant, while the insects' life cycle and behaviors are brought forward in her writings. Merian's insects were depicted in the plates as life-size and as nearly as possible this holds true for the plants, often resulting in the portrayal of only a small portion of plant structure. For the dandelion and plantain, she includes the complete aboveground plant, but she does not show the roots as is done in some contemporaneous herbals, even when she mentions larvae of the cockchafer beetle feeding on them below ground.[32]

The choice of which plant to represent in each plate was another decision, and Merian makes her reasoning transparent through her text. In the 1679 *Raupen* book, she usually anchors the plate with the specific type of plant on which the caterpillars most often feed (the larval host plant), a priority that she makes clear in the book's title. However, she deviates from this pattern in a few places either because there was no room for the plant on the plate (e.g. Plate 17), or she did not wish to repeat the same or similar plants too many times. Some plants are shown twice, once in bloom and in a later plate laden with fruit, but she used the opportunities granted by caterpillars that feed on a wide variety of foods to include more exotic or striking plants, such as the carnation (Chapter 49), sweet sultan (Chapter 46), and the wallflower (Chapter 12). In cases where she depicted insects that typically specialize on only one type of food plant, Merian was careful to show it on this particular host species.

Another set of decisions that she made was the selection of insects to include in her first caterpillar book. The 1679 *Raupen* book featured moths in 38 chapters, although four of these species were pictured twice. She also wrote about two types of beetles and six different butterflies, as well as a variety of gall insects and a number of flies and wasps, many of which parasitize lepidopterans.

---

32   Merian, 1679. *Raupen*, Chapter 4. *Melolontha melolontha* (cockchafer). *Taraxacum officinale* (dandelion) is featured in Chapter 8 and *Plantago major* (greater plantain) in Chapter 36.

These choices were fundamental to the content of the book's images, because insect choice usually dictated plant choice. She included only moths and butterflies for which she had obtained the caterpillar, pupa and adult, but eggs were not a requirement for her images. Another selection consideration might have been whether caterpillars were especially damaging to plants of particular interest to gardeners, or if the species was prevalent in the landscape.

It might reasonably be supposed that her 1679 book included the first species that she successfully raised through metamorphosis. Whether her study journal can help decode how chronology dictated what went into these choices is hard to say. Sixty percent of insects that appear in the 1679 *Raupen* book are found within the first thirty study journal entries. However, within these thirty entries, many of which describe two types of insects, there are several species that appear in the 1683 *Raupen*. Thus, insects included in the first and second *Raupen* books are intermingled in the study journal, and there is no clear-cut point at which the studies for one book end and those for the next begin. As early as entry 35, we begin to see notes and studies for the third caterpillar book, which would not be published until 1717,[33] and these increase in frequency as the number of references to insects from the first two *Raupen* volumes begin to taper off. Evidently, she did not bias her selection of publishable images toward the most colorful species of adult moths and butterflies. Some of the most striking and certainly the largest European lepidopterans are in the family Sphingidae, the sphinx- or hawk-moths, and examples of these do not make an appearance until the 1683 volume of Merian's work (see study journal image in Figure 48). Perhaps by 1679 she had not yet been successful in getting an adult sphinx moth by starting with a caterpillar. None of the study journal entries for the sphinx moths that appear in the 1683 *Raupen* book are dated by year, so it is not possible to know when she reared them.

Composing the Images

The images of caterpillars munching their way along leaves, moths laying eggs, and butterflies about to land on a blossoming plant are hallmarks of Merian's oeuvre. Once she had decided which insects and plants to include in her book, she made a multitude of additional choices in designing compositions that were both informative and pleasing to the eye. The contents of the finished *Raupen* book evoke the passage of time in two ways. The sequential appearance of the book's plants follows the growing season in Germany, where early

---

33   Merian, 1717. *Der Rupsen.*

FIGURE 48   Merian. Study journal painting 33. *Smerinthus ocellatus* (hawk moth) and
              parasitoid

spring tulips, lilacs and flowering cherry trees are followed by buttercups, roses
and gooseberries, and autumn is heralded in the last plate by an acorn-laden
oak branch. The plants are revealed in only one stage of growth at a time, and
in several cases, a common tree or shrub is shown first in flower and later with
fruit. Merian's insects track the seasons as well, but for these she took liberties
with the timing of events. In order to save space and to be clear about the type
of insects she was showing, Merian, like Goedaert before her, condensed weeks
or months of an animal's life cycle into one frame. Anyone reading her text
would not be misled by this "time-lapse" image, for Merian includes the timing
of events in the lives of her subjects, typically stating the month in which she
found the larvae, when they pupated and how much time passed before an
adult emerged. Other evidence of elapsing time appears as tiny larvae hatching from eggs (Plate 1 of the silkworm moth) and caterpillars of increasing size
within one plate, often showing the exoskeletons that are cast off as the larvae
grow. The passage of time also is indicated by leaves that have been chewed,
and details such as the peacock butterfly emerging from its cocoon, as another
cocoon hangs nearby.

   Action is simulated in Merian's nature dramas in a number of ways, most
often by insects pictured as feeding upon their plant hosts. Caterpillars are
seen making their move from one part of a plant to another in several plates

(e.g. Plates 29 and 50), and in these poses the viewer gets a sense of their mode of locomotion. In the composition of several plates, insects are interacting, and although she did not understand the phenomena until much later, she accurately depicted the mortal attacks on caterpillars and larvae by flies and wasps that feed on the living bodies of lepidopterans.

Adult moths and butterflies are posed on plants in about half of her plates, and the other half show them in flight, or with a further sense of immediacy, just as they are about to land upon a plant (e.g. Plates 31 and 47). This device also takes advantage of the fact that most of the adults were preserved with their wings spread. In all of Merian's works on insects, moths are typically shown from above in order to display their wing coloration. Simulated flight is an effective means of doing this, while adding a dynamic element to the compositions. Butterfly wings are often patterned and colored very differently on the top and on the bottom, and for all but the swallowtail Merian shows butterflies in two distinct positions to demonstrate these important differences in the two sides of the wings. The swallowtail butterfly's wings are colored much the same top and bottom, so she simplified that composition to show only a perched adult. By contrast, very few of Merian's moths are shown in two positions. Exceptions to this are seen in Plates 7 and 24, in which she varies wing position of a species of moth, showing them both flying and having landed on a plant. A top view of a moth's wings usually suffices for identification, because moths do not fold their wings upward when they are stationary as butterflies do.

Pupae are generally static, but depicting them presented Merian with a different set of challenges. The pupa of a butterfly, also called a chrysalis, has no silken cocoon around it. The chrysalis usually is attached to a surface, often on part of a plant, and this is how Merian pictures them in her books. Many insects pupate hidden in soil or leaf litter, but she developed no pictorial device for this. She instead positions these stages on or near the host plant of the caterpillar. This space-saving device in her images is neither naturalistic nor accurate, and it led to criticism of her work. However, her text does not say this is the natural location of these pupae. Instead, she usually remarks that she has "shown" them in such a way, not that she found them in this position. Because her subjects transformed from caterpillars into pupae as captives in wooden boxes, Merian had few opportunities to observe their natural pupation sites. Moths that pupate above ground often add the protection of a silken cocoon that surrounds the hardened pupae. For these insects, she uses a variety of devices to show the internal and external structure, sometimes depicting the pupa and cocoon separately. Plate 1 of the silkworm life cycle contains the most detailed representation of the layers of pupa and cocoon, showing four views that make the structures clear. In Plate 1, the shed exoskeleton of

FIGURE 49   Merian. Study journal painting 23. *Rhyparia purpurata* (purple tiger moth)

a silkworm caterpillar is shown lying just above the mulberry leaf. In several chapters, Merian informs her readers through text and image that the last exoskeleton that splits from around the mature caterpillar remains attached to the pupa (see Figure 49). Her choices for illustrating various pupae indicate that she wanted to provide her readers a clear view of the structures involved in what she must have perceived as the most mysterious stage of lepidopteran metamorphosis.

The choices she made led to a finished composition for each plate that incorporated the elements she thought most important to depict. Despite what has been written about the relationship of the images in her study journal to the insects pictured in the published book, there is no evidence that Merian traced the study paintings of insects as she arranged them around the caterpillars' food plants. Nor did she copy each study precisely, slavishly using the study journal as the final arbiter of insect portrayals in her plates. Years often elapsed between the creation of the initial study image and the final plate, and although the studies were her guide, she was usually careful to create a final image that was as true to life as she could make it. Leaving aside the question of coloration, a large number of differences, albeit small in many cases, exist between her original studies and the insects seen by her readers in the printed plates. Her first plate depicting the silkworm life cycle has a number of elements that do not appear in the corresponding study painting, notably eggs and tiny hatching caterpillars. The only conclusion to be drawn from numerous differences between her studies of insects and their appearance in the finished plates is that Merian redrew the insects. Her observational experience, coupled with her artistic skills, likely would have made this easier than trying to trace her originals from the study journal. Departures from the study journal do not diminish its importance to Merian as a source document. Like other scientists before and after her, she recorded her immediate impressions, and for creating her final images, she relied on her studies as a record of size, coloration and in many instances, posture of insects. As she continued to observe insects over the years, she accumulated more visual knowledge, which can be seen in some of the changes made from study to final image. In rendering the insects onto the plates, she often improved upon the original painted study. A clear example of this can be seen when one compares the appearance of the segmented legs of the adult garden tiger moth in Plate 5 (see p. 168) to the less specific leg structure she painted in the study (study journal entry 25, see Figure 64 in commentary). Throughout her career, Merian used her studies as the foundation for her plates, but she added to and subtracted from these earlier images, refining her portrayal of the insects and ultimately combining them with ecologically related plants to create her unique compositions. Once she had composed her final image, she put this design onto a copper plate to be used for making the prints.

Making the Plates

All of Merian's books on European insects had images printed from copper plates onto paper, which when cut and bound resulted in a quarto volume. She asserts that she made her own plates, writing that "there is nothing to be found in this, my modest little book, that was not first painted and then etched or engraved in copper by me...."[34] In her 1679 *Raupen* book she refers to these copper plates a dozen times as if to emphasize the fact the she did the intaglio work herself. Merian also mentions in five places within the book that she first painted the images herself, although it is unclear whether she is referring to the finished compositions or separate studies of the insects and plants. She

---

34   Merian, 1679. *Raupen*, Chapter 3. Merian's reference to her technique reads "*aufs Kupfer geradirt oder gestochen ...*" She uses the phrase *ins Kupfer gestochen* (engraved into copper) on her title page, but on p. iv of her preface she uses the term *radiren*, which refers to the use of a needle. Taken together, her terms seem to suggest that she used some combination of etching and engraving, but this may be more a matter of language than technique.

probably concentrated on the time-consuming work of creating the finished plates for the book from late summer through early spring each year of the project. By the end of August, the number of insects needing her daily attention would have dwindled, and by October, most of her plants and insects would have become dormant for six months or longer.

Despite her use of the term for engraving, all of the plates in her caterpillar books are etched, although Merian may have used engraving techniques to sharpen some lines. She could have learned etching and engraving as a girl or young woman assisting in the family business in Frankfurt, and she may have trained further with her husband. Both of these intaglio techniques result in a copper plate incised with grooves to hold the ink that will make the printed image, but the processes are very different, and to the trained eye, engraved lines usually can be distinguished from those that are etched. Etching requires that the polished copper plate be covered evenly with an acid-resistant ground such as wax or resin, which are often combined. This mixture is melted onto the warm plate, and the design is drawn into the ground with an etching needle, exposing areas of the copper as thin lines.[35]

Merian may have made watercolor paintings of insects and plants together to serve as models for her printed work, and these initial compositions would have helped her to achieve the spatial relationships and aesthetic statement that she desired in the finished print. She usually painted on vellum, and such paintings would serve nicely as long-lasting documentation of her compositions, complete with the colors of the organisms. However, making such a time-consuming and expensive model for her images was neither necessary nor the best tool for transferring an image to a copper plate, and there is no compelling evidence that such paintings were made. An uncolored drawing on paper is much more malleable for tracing onto copper than a painting on vellum, is quicker to make, and less costly. If she did transfer images by tracing, she could have used transparent paper to trace the lines of a painting onto an intermediate transfer drawing. After placing the transfer drawing onto the coated copper plate, the lines could again be traced into etching ground with a sharp tool. Alternatively, a trained and experienced artist like Merian can use a

---

35   For an excellent history of the techniques of etching, engraving and printing discussed here, and further detail on methods, see Stijnman, Ad. 2012. *Engraving and Etching 1400–2000: a history of the development of manual Intaglio printmaking processes.* Houten, Netherlands: Archetype Publications in association with Hes & De Graaf Publishers: 196–209. Ad Stijnman has examined Merian's plates in the 1679 *Raupen* book and says that they were etched (personal communication); I defer to his expert opinion on this.

painting as a model for making a freehand drawing directly on the ground laid upon a copper plate prepared for etching.[36]

The next step in making an etching involves careful application of an acid (e.g. vinegar or nitric acid), which bites into the exposed metal wherever the ground was removed by the etching needle. The design is thus "etched" onto the plate by the acid rather than being carved into the metal by a tool. Lines created by acid etching can be blurred at times, and an engraver's tool, the burin, may be used to dig into the metal to clean up and sharpen these lines, or to add more incised areas. The motion for etching into a ground with a needle is the same as for drawing and requires no specialized training. On the other hand, engraving into metal with a burin requires a steady hand pushing the tool forward, a skill that requires diligent practice and a degree of physical strength.

After the intaglio process is finished, the copper plate was positioned in a roller press made specifically for printmaking. Once the plate is inked and the paper in place, the press was operated by hand to generate the prints. By the 17th century, most plate printing in Europe was done in shops that specialized in this, and not by the engravers or etchers themselves. There were exceptions to this however, and 17th century Nuremberg did not strictly regulate plate printing.[37] If this costly step of the process took place in the Graff home workshop, there would have been further reduction in the production costs of the book, in addition to the considerable amount saved by Merian's work in creating the copper plates. The Graff family most likely had a press for making plates, and evidence for this lies in the existence of counterproof prints from the 1679 *Raupen* book.

Merian's Counterproofs

Counterproof printing was known in Nuremberg at least a century before Merian's arrival in this printmaking center, and she may have learned of it there, or perhaps discovered the process independently.[38] The procedure is relatively straightforward; a print is made in the usual way using paper and

---

36   "Copying a drawn image onto a copperplate, either for engraving or for etching, is daily business for a professional such as Merian." From Ad Stijnman (personal communication April 2018). See also Stijnman, *Engraving and Etching*: 154–9 on the processes used for image transfer to the ground on a copper plate. The preparation of materials such as the ground and acid are complex and highly variable, and each workshop had its own special techniques and mixtures.

37   Stijnman, *Engraving and Etching*: 81.

38   Roth, Michael, Magdalena Bushart, and M. Sonnabend. 2017. *Maria Sibylla Merian und die Tradition des Blumenbildes*. Munich: Hirmer Verlag: 163.

FIGURE 50   Merian, 1679. *Raupen*. On the left is a detail from Plate 44 of *Aglais urticae* (small
             tortoiseshell) as it was typically printed. On the right is the same detail from a
             counterproof print of Plate 44. This mirror image print has more faintly inked lines that
             can more easily be masked with paint to create the appearance of a watercolor painting.

a copper plate, and while the ink is still wet, a fresh sheet of paper is placed
upon the first print and the press is applied again. A print made from a copper plate is the mirror image of the plate, and this image will be reversed
once again in the counterproof. As a result, the counterproof is in the same
orientation as the original drawing or painting. Counterproofs have two
other attributes useful to the artisan, which is that their production does not
wear out the copper plate, and the outline of the plate is not pressed into the
counterproof support as it is with the usual type of print. If a counterproof is
made with care, the inked lines may be very faint, and an artist can then paint
this image, producing a work that looks very much like an original painting.
Comparison of a printed plate to a counterproof print shows that the images
are reversed and that the printed lines of the latter are much fainter. If the
artist who colors the counterproof uses body color rather than watercolor, the
printed lines are further obscured (Figure 50). Because counterproofs must
be made from a freshly inked print, the maker must have access to a press
and the original copper plate of the image. Merian made dozens of counterproofs in her career, and this fact has been used in support of the idea that
she had access to a press and possibly had a hand in printing the plates for her
books.[39] At least one unpainted counterproof copy of the 1679 *Raupen* book

39   Merian, Van Delft, and Mulder, *Merian Metamorphosis*: 43 and Lölhöffel, 2015. Maria
      Sibylla Merianin und Johann Andreas Graff, Erster Teil: 56. When her daughters were old
      enough to assist in the family workshop, they were likely employed in this process as well.

exists today, so in this case there was no attempt to portray this volume as having original paintings in it.[40]

The counterproof edition of the plates used in the present book are examples of some thought to have been painted by Merian or in her workshop.[41] Comparison of these counterproof plates with her study journal images shows very similar coloring and therefore lends credence to this idea (see for example Plate 1 and Figure 61). The counterproof process continued with later volumes of her work, including *Metamorphosis* and the Dutch *Rupsen* (caterpillar) books. In addition to the creation of counterproof prints on paper, which was not uncommon, Merian and perhaps others in her workshop (i.e. her daughters) created counterproofs on vellum, something that seems to have been more unusual among artists in her time. A number of watercolor paintings on vellum existing in documented collections match the *Raupen* book's plates, and some of these were once thought to have been the "original" paintings that served as models for her plates. Further examination of most of these works has shown that the majority are painted counterproofs, created after the copper plates were made.[42] The small number of existing watercolors matching the *Raupen* plates that are not counterproofs could also have been painted after the plates were printed; neither these nor the counterproofs are dated, and the provenance of most of these works is unclear.

Lastly, we must consider how the existing hand-colored plates in copies of the *Raupen* book came into being. Merian offered both uncolored and colored editions of her books for sale, and described these various offerings toward the end of her 1679 preface.[43] The colored editions would have been much more costly; hence, most copies were colored post-purchase and not by Merian. As best she could, she used her text to describe the colors of all stages of the insects, but there are many variations within green, yellow, brown and even in shades of white. People with a wide range of skills colored the plates, and some of them evidently had little understanding of the colors of the organisms in nature. This tremendous variability in the quality and naturalism of hand-coloring in the *Raupen* plates is not unique to Merian's work, but is widespread among illustrated books hand-colored after they were sold (sometimes decades later). Because there is typically no indication as to who colored the plates in a book,

40   This unusual uncolored counterproof volume is in the Trew Collection within University Library of Erlangen-Nuremberg. Shelfmark H61/4 TREW.P 73/74.

41   See pp. 129–30 for more on this specific counterproof edition.

42   Roth, Bushart and Sonnabend, *Tradition des Blumenbildes*: 163–5.

43   Merian, 1679. *Raupen*: preface. "But if the nature- and art-loving reader wishes to have the entire plates, or only the caterpillars and their changes, along with the insects, accurately colored, we will gladly supply either one."

subsequent readers familiar with an organism's typical coloration might attribute deviations in coloration to errors or carelessness by the original author. Such false impressions did occur with respect to Merian's work, and in some instances, the mistakes of others may have unfairly damaged her reputation as a naturalist. However, her accurate eye and ecological compositions were very well received, and in the decades to come, much emulated by others who produced illustrated natural history books.

Composing the Text

In addition to working on the copper plates, Merian could have used her time in the colder months to organize her notes and to write the text for her book. What is certain is that the text was finished after the plates, because positional information is included to point her readers to features of the composition, such as a caterpillar "placed at the top, on the rosebud."[44] As with her other caterpillar books, the 1679 *Raupen* featured an illustrated title page (Figure 60),[45] a letterpress printed title page, fifty numbered plates, and the accompanying text. In addition to fifty short chapters corresponding to fifty plates, she included introductory text in her preface. The book also has two indices to assist her readers in finding insects and plants in the book (one index for common names and an index of Latin plant names). The first chapter on silkworms is the longest at four pages each, and the subsequent chapters are limited to two pages each. The organization of the chapters is roughly chronological and follows the growing season in Nuremberg. Merian notes some exceptions to this plan but states that:

> ... all the other caterpillars in this little book follow one another according to when, from month to month, they were accustomed to spinning themselves up. The same is true, in almost every case, of the flowers and other plants available for them to feed on at such times....[46]

In each two-page "chapter," Merian recounts in a straightforward manner the stages by which each insect passes through its metamorphosis, usually

---

44 Ibid., Chapter 19. Plate 19 represents two species of Tortricid moth and their respective caterpillars. Merian indicates the position of each species in the finished plate.

45 This type of plate typically would be cataloged as an "engraved title page," but Merian's image of the mulberry wreath and insects was etched; the text may have been engraved. To avoid confusion, I will refer to this plate as the illustrated title page.

46 Merian, 1679. *Raupen*, Chapter 34.

mentioning the time of year and duration of pupation. When parasites and other insects are noted, she discusses them as well as her primary subjects, moths and butterflies. Text entries vary in how closely they follow the notes in her study journal, ranging from almost complete congruence to much expanded descriptions in the finished book. She included information about the pictured plant, sometimes naming it as a preferred food of the caterpillars, but in other cases adding to what she recorded in her study notes, perhaps giving the plant's local name, or explaining how insects affect the plant. Another difference between the study journal notes and her finished text is the inclusion of details of insect coloration in the latter. Merian had her own watercolor studies for reference and would not have needed to include description of pigmentation in her written research notes. Anyone buying an uncolored copy of the *Raupen* book would have needed color references to have more complete information on the appearance of the insects at each stage. To aid her readers, she had the printer change the font size or weight to emphasize the words for colors and other physical descriptors that applied to the insects; this font size change was also used to emphasize the names of plants, and in a few instances, insect names or her terms for them. For example, an excerpt from p. 12 of the 1679 *Raupen* reads:

> For fourteen days they remain like this, undergoing their change, after which the MOTTLED ADULT MOTH at the top emerges, a lovely BRIGHT RED marked with BLACK SPOTS and STRIPES on its BODY and HIND WINGS, but with A BROWN HEAD AND FOREWINGS, which are also marked with WHITE STRIPS. They have **six** RED LEGS and TWO BROWN-AND-WHITE HORNS; and they produced PALE SEA-GREEN EGGS resembling grains of millet.

Interestingly, bold type is used for such emphasis of terms in the preface, but an increase in font size is used for this purpose in the subsequent chapters.[47]

The use of first person in her writing allows Merian to insert the occasional personal remark, to express religious sentiments, and to let her joy in the discovery shine through. Some comments include a glimpse of her ambitions: "... if it is your wish, I will pursue this work in service to my readers, that I may keep their interest keen with art and earn the praise and favor of great men." Other comments hint at her personal struggles, such as the remark that she was "obliged to compile it [her book] while managing her household."[48]

---

47   Ibid., Chapter 5. We have indicated the emphasized words in capital letters.
48   Ibid., Chapter 50.

Throughout the book, Merian takes us along with her as she explores the world of caterpillars, moths and butterflies, and we have a sense of her curiosity and pleasure as she writes of "thought-provoking and elegant transformations."[49] A modern reader might find her means of expression at odds with the scientific content, but her tone and style of writing was not unusual for naturalists at the time.

A substantive addition to her finished text, absent from her study journal, is the inclusion of generalized conclusions drawn from years of observations, such as the differences between what she terms "summerbirds" (butterflies) and "moth-birds" (moths). Her statements on the effect of temperature and rainfall on her small subjects, their food choices and some general behavior patterns also result from long-term investigation of her subjects. Thus, much in terms of ecological content is not in her study journal but is central to the published text. Yet Merian did not attempt to explain everything that she observed, and with a few exceptions, she avoided speculation, which sets her apart in another way from many naturalists of the period. The only part of the text not written by Merian are two poems by Christoph Arnold, both of which take a very different tone from her chapters. As mentioned earlier, she and Arnold were acquainted, but no record of any correspondence by letter with Arnold is known, and how he came to contribute to the volume is unclear. Either she invited his contribution or he volunteered it, but in any case, she must have welcomed the additions by the notable Nuremberg scholar. Such laudatory offerings were common to books at the time, but even so, Arnold's "Song of Praise" was strong validation of her industry and ingenuity.

Financing, Printing and Marketing the *Raupen* Book

As discussed in the first chapter, the cost of producing a heavily illustrated book was an obstacle that many naturalists were unable to overcome in order to publish their work. Merian's training in intaglio techniques likely contributed in large part to her ability to produce her flower and caterpillar books at a relatively young age. Because she researched, wrote and illustrated these works, the primary labor costs went toward the printing. If, as I propose, she and her husband had a hand in printing the plates, this expense would have been reduced to the sum paid for printing the text. Some authors attempted to sell subscriptions in advance of publication in order to defray the heavy costs of illustrations; Merian followed this path for *Metamorphosis*, for which she

---

49   Ibid., Chapter 27.

hired craftspeople to etch the plates,[50] but there is no record showing that she sold subscriptions for the 1679 *Raupen* volume. By producing the European books as quarto volumes, the cost of the paper and copper plates would have been much less than for the folio size of *Metamorphosis*. The octavo volumes produced by Goedaert for his books on insects were even smaller and would have been less expensive to print than the quarto *Raupen* volumes, requiring less copper and paper. However, Merian needed the larger quarto format to allow space for inclusion of the caterpillars' food plants in the plates, while still maintaining the life-size scale of the insects. Although Merian's labors cut the cost of the book's production by at least half, copper plates and paper had to be procured and the text printed by a professional workshop. Several possibilities exist for the means by which she financed these remaining expenses of publication. Merian earned money teaching young women to paint, selling her own artwork, and by selling artist's pigments. She also would have profited from the sale of her first two sets of flower prints (the 1675 and 1677 *Blumen* books). It may be that family money (hers or Graff's) helped to defray costs until her first books could be sold.

The paper used for printing the 1679 *Raupen* book has not been professionally analyzed, but it appears to be of high quality and has stood the test of time. Unlike the plates, which were printed on one side of a single sheet of paper, the text was printed on both sides of a sheet. The complete publication, like most of its time, would have been sold as separate sheets, and when bound by the new owner would have resulted in a book in which almost half of the sheets were made up of the plates. Merian designed her book such that the short chapters on the two sides of a page could be bound adjacent to the accompanying plate for easy referral to word and image. As with most books of its kind, the paper for the plates was slightly heavier than that used for the text, in order for the paper to withstand the more intense pressures of the roller press used for plate printing.

Printing of text required different skills and equipment than the printing of plates. The text for Merian's 1679 *Raupen* book was printed in Nuremberg by Andreas Knortz. For unknown reasons, she used a different Nuremberg-based

---

50   Merian was ill when she returned from Surinam, and the task of etching and engraving all 60 folio size plates while writing the text and marketing the book may have been difficult at this point in her life. In order to be able to afford the best artisans to etch the plates, she wrote to people in her network of naturalists and collectors soliciting subscriptions from parties interested in buying the book in sections or as a whole when finished, a common practice at the time. Some of these letters have been published in Merian and Wettengl. *Maria Sibylla Merian*: 264–7. For more on the artisans employed by Merian to etch plates for *Metamorphosis*, see Merian, Van Delft, and Mulder, *Merian Metamorphosis*: 46. She or one of her daughters may have also contributed plates to the 1705 book, as three are unsigned.

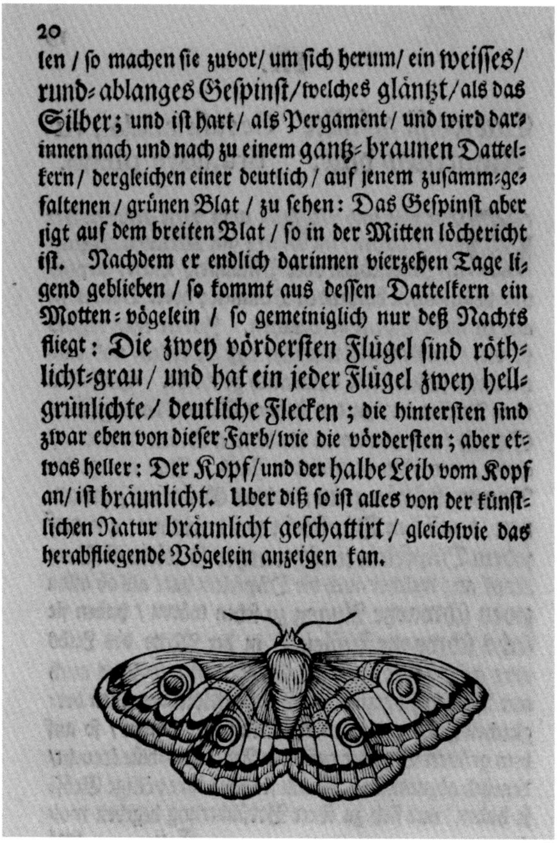

20

len / so machen sie zuvor/ um sich herum/ ein weisses/ rund-ablanges Gespinst/welches gläntzt/als das Silber; und ist hart/ als Pergament/ und wird dar-innen nach und nach zu einem gantz-braunen Dattel-kern/ dergleichen einer deutlich/ auf jenem zusamm-ge-faltenen/ grünen Blat / zu sehen: Das Gespinst aber jigt auf dem breiten Blat / so in der Mitten löcherigt ist. Nachdem er endlich darinnen vierzehen Tage li-gend geblieben / so kommt aus dessen Dattelkern ein Motten-vögelein / so gemeiniglich nur deß Nachts fliegt: Die zwey vörderſten Flügel sind röth-licht-grau / und hat ein jeder Flügel zwey hell-grünlichte / deutliche Flecken ; die hinterſten sind zwar eben von dieser Farb/wie die vörderſten ; aber et-was heller: Der Kopf/und der halbe Leib vom Kopf an/ ist bräunlicht. Uber diß so ist alles von der künſt-lichen Natur bräunlicht geschattirt / gleichwie das herabfliegende Vögelein anzeigen kan.

FIGURE 51
Page 20 in Merian's 1679 *Raupen* book features a printer's ornament showing an emperor moth, a species she pictured in Chapters 13 and 23.

printer, Michael Spörlin, for the text of the 1683 *Raupen* book. In both *Raupen* volumes, Merian's creative control can be seen even on the printed pages. It was common practice to use printers' 'ornaments' to fill the bottom of a page when the type did not stretch to the bottom of the printing frame (Figure 51). The 1679 *Raupen* book has seven versions of these small decorative figures. Although printers had such ornaments on hand as tools of their trade, I believe that Merian designed at least five of the ones used in her *Raupen* books: a rose with chewed leaves adorned by two moths, an anemone, a tulip, a small moth, and a larger and more elaborate emperor moth (Figure 52). The rose, anemone, tulip and both moth ornaments were used again in the 1683 *Raupen* volume, and the rose appears as the sole print-er's ornament in the 1680 *Blumenbuch*. The ornaments may be woodcuts, which could be printed on a page of text with the same type of press used for type. Because she employed a different printer for the 1683 publication, the re-use of these five ornaments is evidence that the designs were created by Merian.

FIGURE 52   Five printer's ornaments that were used several times each in Merian's 1679
              and 1683 *Raupen* books. The damaged lower petals on the rose and the
              asymmetrical flowers reflect Merian's style, and the moths are similar to some
              appearing in her plates. The approximate relative scale of these ornaments is
              shown here; the larger ones were used to fill larger blank areas on a page. From
              Merian, 1679. *Raupen*.

The role of Johann Andreas Graff in the publication of the 1679 *Raupen* book
is unclear, and neither this book's title page nor those of the *Blumen* books
help to illuminate how the couple may or may not have collaborated on these
works.[51] Some scholars have named Johann Andreas Graff as the publisher of
Merian's *Raupen* books, but if this was the case, it is odd that the title page
would not have made this evident, rather than boldly stating that the work was
"personally published by Maria Sibylla Graff."[52] As described earlier, Johann
Andreas had his own successful career in Nuremberg as a draughtsman and
engraver/etcher, and his extensive body of work would have kept him very
busy. However, the industrious Graff family would have sold both her works

51   For further discussion of the front matter for all of Merian's 17th century publications, see
      Pick, Cecilia. 2004. *Rhetoric of the Author Presentation: The Case of Maria Sibylla Merian*.
      Doctoral Dissertation, University of Texas, Austin: 77–82.
52   Publication information on the 1679 title page uses the intensive pronoun form, "*selbst
      verlegt*," which could also be read as "self-published" by Maria Sibylla Graff. As Margot
      Lölhöffel has pointed out, this is "unusual proof of self-confidence and self-determination
      for a married woman at that time," and may be further evidence that 17th century
      Nuremberg offered a fair degree of freedom for married women to pursue work in their
      own name (personal communication, October 2018).

and his from their home in Nuremberg.[53] Maria Sibylla marketed her books in a variety of ways during her lifetime, including print advertisements, letters to friends and collectors, and by receiving visitors who wished to see her collections, artwork and books. She engaged in cross-marketing techniques as well, promoting the 1679 *Raupen* book in her 1680 *Blumen* book preface. In order to establish the validity of her work, she kept insect specimens for display, writing that "... I continued to observe them into the fifth year and discovered astonishing changes, keeping them all in boxes and showing them to anyone wishing to see them."[54]

The value of the 1679 *Raupen* book as a contribution to the history of entomological research must be judged by the originality and accuracy of its content, which is discussed in my commentary on Merian's text and her images. Understanding her meticulous processes in producing this work provides the background needed to validate her work, but also gives us insight into ways in which her books were unusual relative to most of the earlier books on insects. The lack of intermediaries in the production of the illustrations (e.g. hired draftsmen or engravers) stands in contrast to the production process of many of the natural history books that resulted from the work of her near contemporaries. Her level of involvement not only was the means by which her earliest books were cost effective to produce and sell, but also gave her an unusual degree of creative license and a great deal of control of her finished product. She employed this control to vary the structure and content of her images in order to maintain viewer interest and to attract buyers. In the images from the 1679 *Raupen* reproduced in the present book, one can see that she has created a series of small natural dramas, where time passes, movement by insects on plants and in the air is suggested, and destruction and death by herbivory, predation and parasitism take place. At the same time, new life appears, growth and reproduction occur, and the seeming miracle of metamorphosis plays a central role in events.

---

53 A 2017 exhibition in the Nuremberg City Library presented a subsample of Graff's extensive works; these featured large and highly detailed prints of cityscapes and architectural marvels in and around Nuremberg. Colditz-Heusl, Silke, Margot Lölhöffel, Theo Noll, and Werner Schultheiss. 2017. *Katalog zur Ausstellung: Johann Andreas Graff, Pionier Nürnberger Stadtansichten im Kunstkabinett der Stadtbibliothek Nürnberg*. Nuremberg: Association of Friends of the Nuremberg Cultural History Museum.

54 Merian, 1679. *Raupen*: Preface.

CHAPTER 4

For the Benefit of Naturalists

The 1679 *Raupen* book was Maria Sibylla Merian's first scientific publication,
but far from her last. Tracing the history of her full body of work is useful in
understanding her influence on subsequent biological inquiry and on her reputation as a naturalist. The first *Raupen* volume set the pattern for her later
books, and in 1683, she published a second caterpillar book.[1] Merian's research
on insects continued for almost three more decades after the appearance of
the second *Raupen* volume, but publication of the third volume on European
insects was put on hold until it was published in Dutch in 1717.[2] In the interim,
she researched and wrote about Neotropical insects and plants, and in 1705 she
published her best known work, *Metamorphosis Insectorum Surinamensium*.[3]
In the years between the Surinam volume and her 1717 *Rupsen* book, the earlier
German caterpillar books were translated and published in Dutch. In the 18th
century, all of Merian's books were published in a variety of editions in Dutch,
Latin and French. This somewhat complex history is noteworthy, because edi-
torial decisions by Merian and by publishers who controlled her material after
her death had a lasting effect on how she has been perceived as a scientist.

The 1683 *Raupen* book contained additional metamorphoses that Merian
studied before moving from Nuremberg. Although this second volume was
similar in format to the first caterpillar book, there are differences in how she
arranged the content. As she states in the preface to the second volume, "more
than a hundred transformations can be found here …," and she tells readers
that for their convenience, she has combined insects that eat the same plant
into a single plate. She also explains that she has added "graceful flowers instead of the various foods [plants] which have already appeared several times
in my [1679] book." The second *Raupen* book does have more insects depicted
than the first volume and the expanded plant array includes fewer fruit trees.
Some of the flowers depicted in the 1683 volume would have been popular in
gardens at the time, but others are more humble meadow plants. As in the first
volume, she was careful to denote the food plants of her caterpillar subjects in
the text, even if she posed the insects on a different species.

---

1 Merian, 1683. *Raupen.*
2 Merian, 1717. *Rupsen.*
3 Merian, *Metamorphosis.*

© KONINKLIJKE BRILL NV, LEIDEN, 2021 | DOI:10.1163/9789004284807_005

The focus of the present book is on Merian's early investigations into European insects, but no study of her work is complete without at least a brief discussion of *Metamorphosis*.[4] Intrigued by the tropical moths and butterflies from the Americas that she saw in the collections of others, she and her youngest daughter, Dorothea Maria, set out for Surinam in 1699. She worked there for almost two years before becoming seriously ill and returning to Amsterdam. That city had been her home since 1691, when she left the Labadist community in Friesland. The contents as well as the physical size of the Surinam volume departed dramatically from the *Raupen* books, as befitted the subject. As with all of her insect images, the tropical species featured in *Metamorphosis* were reproduced as life-size, requiring much bigger plates to accommodate the enormous wingspan of the largest insects,[5] and to present a fuller picture of the tropical plants on which they lived. Merian understood that her audience would be intrigued by the exotic plants, and she addressed these in almost as much detail as she did the insects. The vast array of tropical insects were strangers to her when she arrived in the hot, humid forests of Surinam, and she had a very short time in which to learn about the life cycles of dozens of new species described in her book. It is ironic that although *Metamorphosis* is considered her *magnum opus*, the biology of insects is more accurately described in the less celebrated *Raupen* books.[6] *Metamorphosis* initially was published in Dutch and Latin, and later in French. The extraordinary visual presentation of striking plants and animals in the Surinam book brought a new surge of fame to Merian, and ultimately led to further dissemination of the contents of her earlier work.

Twelve years after the publication of *Metamorphosis* and shortly after her death, Merian's final caterpillar book was published by her daughter, Dorothea Maria.[7] This third book on European insects was a quarto volume similar in format to the earlier *Raupen* books, but was published in Dutch. Merian investigated some of the insects featured in the 1717 *Rupsen* book while she was still living in Nuremberg; others were observed while she was living in Friesland or later in Amsterdam. As will be described below, she had earlier published

---

4   Merian, *Metamorphosis*. The history of this, her best-known work, has been well documented elsewhere. See for example Merian, Van Delft, and Mulder, *Merian Metamorphosis*; Merian and Wettengl, *Maria Sibylla Merian*; Merian, Maria Sibylla, Elizabeth Rücker, and William T. Stearn. 1980–1982. *Metamorphosis Insectorum Surinamensium*. London: Prion.

5   *Thysania agrippina*, the white witch moth in Plate 20 of *Metamorphosis*, can have a wingspan of 30 cm (almost 12 inches).

6   Etheridge, Kay. 2016. The biology of *Metamorphosis Insectorum Surinamensium*. In: Merian, Van Delft, and Mulder, *Merian Metamorphosis*: 29–39.

7   Merian, 1717 *Rupsen*.

Dutch translations of the 1679 and 1683 *Raupen* volumes, and in 1718, all three *Rupsen* books were published in a single volume in Latin.[8] The Dutch *Rupsen* books and the Latin editions merit closer attention, because these volumes had a significant effect on Merian's reputation in the years that followed.

The *Rupsen* Books

From their inception until after the publication of *Metamorphosis* in 1705, the 1679 and 1683 *Raupen* volumes were available only in German. However, Merian was an astute businessperson, and when her supply of printed copies of the *Raupen* books was depleted, she took advantage of the burgeoning interest in natural history in the Dutch Republic and had the earlier caterpillar books translated into Dutch. The copper plates for the *Raupen* volumes were still in her possession, but rather than printing them as they were, she decided to re-etch several plates in order to insert the metamorphic cycles of several more insects (Figure 53). The added insects are mentioned in her short preface to the first *Rupsen* book:

Beloved Reader

This small book, dealing with 59 observations of insects, I have formerly published at the request of many respected amateurs in Nuremberg. But since the High-German language is not used in this country, I have again been advised to content many amateurs of the bloodless creatures, [and] have brought this to light in the Dutch language, with the addition of fifteen more observations with their descriptions. If only I may have the fortune of pleasing some with this minor work, this will incite me to publish beyond these creatures I have found in Germany, also those that I have collected in Friesland and in Holland in great number. Farewell.[9]

---

8   Merian, Maria Sibylla. [1712]. *Der Rupsen Begin, Voedzel en wonderbaare Verandering.* Volumes 1 and 2. Amsterdam: Maria Sibylla Merian. The Latin edition was comprised of all three of her volumes on European caterpillars and appears to have been translated from the Dutch editions. Merian, Maria Sibylla. 1718. *Erucarum Ortus, Alimentum et paradoxa Metamorphosis....* Amsterdam: Johannes Oosterwijk.

9   Merian, 1712 *Rupsen*: 2. Translation by Florence Pieters and Hans Mulder. The plates in the first *Rupsen* book have insects added to Plates 2, 3, 7, 11, 12, 13, 18, 27, 30, 33, 39, 41 and 50 from the 1679 *Raupen* book. Her motivation for making these additions to the plates was not stated in her writings. See note 49 in Part 2 of this volume for discussion of the term "amateur" as used in Merian's time.

FIGURE 53
Merian. 1712 *Rupsen*, Volume 1.
This counterproof print of
Plate 2 shows added insects,
which were left uncolored in
this volume.

The publication dates were not included in the front matter of the first two *Rupsen* books, and we must attempt to reconstruct a timeline from a few clues. An advertisement for the second volume of the *Rupsen* book (originally the 1683 *Raupen* book) appeared in May 1712, so it is likely that the 1679 *Raupen* book was translated and published in Dutch before this date.[10] We can be certain that the original copper plates of the first caterpillar book were altered before the 1711 visit to Merian by Zacharias Conrad von Uffenbach, because plates in the 1679 *Raupen* book that he purchased from her in Amsterdam include the additional insects described above. Another clue lies in the strange insertion of the *Rupsen* book title on the illustrated title page of von Uffenbach's copy

10   The publication dates that are usually recorded for the first two *Rupsen* books (1713 and
     1714) are incorrect. See: Mulder, Hans. 2014. Merian puzzles. Some remarks on publica-
     tion dates and a portrait by Johannes Thopas, in: *Proceedings of the symposium "Exploring
     M. S. Merian,"* Amsterdam, 2014 on the website of *The Maria Sibylla Merian Society*.
     https://www.themariasibyllameriansociety.humanities.uva.nl/research/essays2014/
     accessed 26 February 2020. Mulder discovered an advertisement in the 31 May 1712 peri-
     odical *Oprechte Haerlemsche Courant* announcing that the second part of M. S. Merian's
     European Insects was available at her house in the Kerkstraat in Amsterdam.

FIGURE 54
Illustrated title page of the
edition of the 1679 *Raupen*
book bought in 1711 from
Merian by Zacharias Conrad
von Uffenbach . For this
printing, her married name
has been removed from the
lower part of the wreath and
Merian is used in lieu of geb:
Merianin. For comparison
see the title page from the
earlier edition shown in
Figure 60.

(Figure 54),[11] which proves that by 1711 Merian also had altered this plate from its appearance in the original *Raupen* book. This change was accomplished by removing the German title from the copper plate and replacing it with the Dutch. The publication of these Dutch editions towards the end of Merian's life raises a number of questions, but here I will focus on how they were different in content from the German *Raupen* books of 1679 and 1683, and will briefly discuss how they may have been assembled.

For the plates of the first two *Rupsen* books that were altered to depict more insects, the accompanying text included concise information on the added species. Even so, the Dutch translations contain roughly one-third to half the text of each of the original chapters in the 1679 and 1683 *Raupen* books.

---

11   The chimeric copy of Merian's 1679 *Raupen* book now in the collection of the Staatsbibliothek Bamberg, Shelfmark HV.Rar.103(1), has other odd features. The letterpress title page was reprinted in German with the same information as the original, but in a different font. The preface is only half the length of the original, and omits most of the personal comments in the original 1679 *Raupen* book. The preface, like the rest of the text, is in German, but has some errors, indicating that it may have been prepared quickly and perhaps by a Dutch printer with limited knowledge of German.

Unfortunately for scholars who used these later editions and for Merian's legacy, much of the behavioral and ecological detail that added such scientific value to the earlier German caterpillar books was omitted from the Dutch editions. Nor does the shortened text convey her personal delight and sense of wonder regarding her small subjects. For most of the entries in the *Rupsen* books, readers are given short physical descriptions of the insects, mention of the timing of metamorphosis and the preferred larval food plants but little other information. Compare for example Chapter 8 from the 1679 *Raupen* book (translated on p. 184 in this book) to this abbreviated translation of the same chapter in the Dutch edition:

> On this wild flower one finds in April a caterpillar with a brown body; it has on its head something like two horns of black hair, and on the back another five such upright standing tufts of hair, and is further covered with yellow hair all over the body. In early May they make an oval cocoon from their own hair, and change into a brown pupa covered with yellow hair, as can be seen below. Towards the end of May a gray small moth-bird came out, as is shown above on a leaf.[12]

Omitted from the Dutch text is the description of the caterpillar's defensive behavior and its habit of making a cocoon from its own hairs mixed with wood that it chews. Nor is there mention of the fact that she found them on hawthorn plants as well as on the pictured dandelion, or that the adult moths fly only at night. This diminution of the original text of Chapter 8 is typical of all of the entries in the *Rupsen* volumes. Also missing from the Dutch editions are Merian's broader observations based on years of study, such as the effects of weather on the growth and survival of caterpillars. It is not known who wrote the revised text, but a strong case may be made for the younger daughter, Dorothea Maria, as the Dutch translator. Her older sister, Johanna Helena, was living in Surinam with her husband during much of this period. Dorothea Maria lived with her mother in Amsterdam while her husband, a ship's physician, was at sea for prolonged periods.[13] She would have been fluent in Dutch by this time, having lived in Friesland at the Labadist colony from the age of seven and in Amsterdam after that.

---

12 Translation by Florence Pieters. Many other differences exist in the tone and content of the German volumes when compared to the shorter Dutch translations, but this would be a lengthy analysis and so is not included here.

13 Davis, *Women on the Margins*: 178. Dorothea Maria married Philip Hendrix within a year of their return from Surinam.

The additions to the plates of the *Rupsen* books tell us about some of the research that Merian carried on after the publication of the German caterpillar books. She was fascinated by galls for decades and curious about what formed these structures and the creatures inside of them. Plate 50 of the first *Rupsen* book was altered from the 1679 original by adding a leaf gall (for which she used the term *galnoot* – gallnut) to the oak branch, along with an image of a gall cut in half (Figure 55). The accompanying text was altered as well, and in this case, intriguing new biological information supplanted much of what was in the original German text, the translation of which is on p. 356 of this book. The Dutch text reads:

> These Caterpillars were green and yellow striped, after having stripped off their skin, they became more brown, and when larger they peeled again and became dark red. I have fed them with these leaves up to September, then they changed into brown pupae from which in December brown, yellow and white-spotted moths emerged.

> On one of the leaves a round node is shown, which is a kind of gallnut. After having opened this in Schwalbach in the Year 1684 in July in the presence of some medical doctors, we found right in the middle a hole, in which lay a small round seed; after twelve days we resumed this observation, and having opened some of these, we found in each Gallnut two holes like a core in an Apple, in each hole laid a little white Worm, but [because] the circumstances required [me] to travel away from there, I could not do research on it further, leaving the rest to other Amateurs.[14]

The gall pictured in the altered *Rupsen* plate was described in her study journal in an entry dated 1684.[15] Because she made this discovery after the publication of her first two *Raupen* books, Merian took the opportunity provided by the re-issue of her work in Dutch to include new information that she considered important. Omitted from this revised final chapter were the personal comments she made in the closing section of the 1679 *Raupen* book.

Regardless of who wrote the shortened text for the *Rupsen* volumes, one must wonder about the reasons for omitting so much of the hard-won information that was included in the German volumes. The primary advantage to the reduced text was that the amount of paper needed for printing was greatly

---

14   Merian, 1712. *Rupsen*, Chapter 50. Translation by Hans Mulder.
15   Appendix, study journal entry 212. One doctor is identified in the journal entry as Dr. Faberitzius [Fabricius], Palatinate physician-in-ordinary and the date given as July 1684.

FIGURE 55
Merian, 1712. *Rupsen*,
Volume 1. This
counterproof print
of Plate 50 shows the
addition of a gall on
the oak leaf to the
lower right and the
cross section of a gall
below this. Compare
to Plate 50 in the 1679
Raupen book on
p. 354 of the present
volume.

reduced; hence, the book could be produced and marketed for sale at a lower cost. The German 1679 edition had 116 pages of text including the front matter, and was printed front and back on fifteen sheets of paper in order to produce a quarto volume.[16] The entire text of the *Rupsen* volume that corresponds to the 1679 *Raupen* book was printed on four sheets of paper. Thus, the cost of paper for the corresponding Dutch volume was reduced by almost seventy-five percent, and there would have been a corresponding decrease in the cost for typesetting and printing the truncated text.

Perhaps Merian did not consider the long-term effects of reducing her text, or if she did, the need to produce the Dutch translations quickly and at a relatively low price overrode these concerns. She may have had debts from the

---

16   Books were printed on full sheets of paper that were folded into quires. Quarto books like
     Merian's caterpillar volumes were assembled from sheets that were folded twice to create
     the quires of eight pages; these were then stitched together.

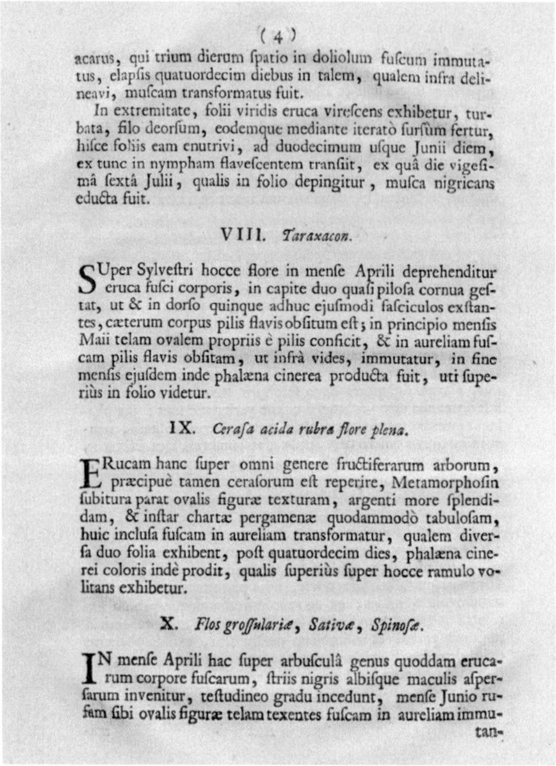

(4)

acarus, qui trium dierum fpatio in doliolum fufcum immuta-
tus, elapfis quatuordecim diebus in talem, qualem infra deli-
neavi, mufcam transformatus fuit.

In extremitate, folii viridis eruca virefcens exhibetur, tur-
bata, filo deorfum, eodemque mediante iteratò furfum fertur,
hifce foliis eam enutrivi, ad duodecimum ufque Junii diem,
ex tunc in nympham flavefcentem tranfiit, ex quâ die vigefi-
mâ fextâ Julii, qualis in folio depingitur, mufca nigricans
eduéta fuit.

VIII. *Taraxacon.*

SUper Sylveftri hocce flore in menfe Aprili deprehenditur
eruca fufci corporis, in capite duo quafi pilofa cornua gef-
tat, ut & in dorfo quinque adhuc ejufmodi fafciculos exftan-
tes, cæterum corpus pilis flavis obfitum eft; in principio menfis
Maii telam ovalem propriis è pilis conficit, & in aureliam fufcam pilis flavis obfitam, ut infrà vides, immutatur, in fine
menfis ejufdem inde phalæna cinerea produéta fuit, uti fuperiùs in folio videtur.

IX. *Cerafa acida rubra flore plena.*

ERucam hanc fuper omni genere fruétiferarum arborum,
præcipuè tamen ceraforum eft reperire, Metamorphofin
fubitura parat ovalis figuræ texturam, argenti more fplendi-
dam, & inftar chartæ pergamenæ quodammodò tabulofam,
huic inclufa fufcam in aureliam transformatur, qualem diver-
fa duo folia exhibent, poft quatuordecim dies, phalæna cine-
rei coloris indè prodit, qualis fuperiùs fuper hocce ramulo vo-
litans exhibetur.

X. *Flos groffulariæ, Sativæ, Spinofæ.*

IN menfe Aprili hac fuper arbufculâ genus quoddam erucarum corpore fufcarum, ftriis nigris albifque maculis afper-
farum invenitur, teftudineo gradu incedunt, menfe Junio ru-
fam fibi ovalis figuræ telam texentes fufcam in aureliam immu-
tan-

FIGURE 56

Merian, 1718. *Erucarum Ortus:* 4. The Latin text of Chapters 8 and 9 from her 1679 *Raupen* book often are less than half as long as they were in the original German.

production of the costly Surinam work if copies of that expensive book were not selling quickly enough. Certainly, the 18th century editions made her work available to a larger audience. However, the textual content that was read by later naturalists made Merian's research seem much less substantial than it actually was. Another disadvantage to readers of these truncated editions was that the text was printed in a way that made it impossible for the plates to be bound sequentially after each entry as they usually were for the earlier German volumes. In the Dutch (and later Latin) editions of Merian's caterpillar books, two or three shortened chapters were printed on one page, so that however the pages were bound, some images inevitably would be separated by two or more pages from the text (Figure 56). In most existing volumes of these later editions, the complete text of each book is bound together and the entire set of plates is bound in after the text. This unfortunate separation of word and image makes Merian's intended integration of the two forms of information more difficult to access than when they are side by side as was typical in the German editions.

The Last Caterpillar Book

The final *Rupsen* book, published in 1717, featured fifty new plates and chapters on the insects depicted in them, thus adding a large number of new European species to Merian's public body of work.[17] Dorothea Maria was acknowledged on the letterpress title page as the publisher of the 1717 *Rupsen* volume, and she wrote the preface, evidence that she was heavily involved in the production of that volume.[18] Although this final caterpillar volume came into print after Merian's death in 1717, it was based on her research, as documented by the small paintings and entries in her study journal. Some entries for insects appearing in the 1717 *Rupsen* are dated as early as 1672 and 1681, and others are based on entries added to the study journal as late as 1710, indicating the ongoing nature of her research on insects. Several images in this volume differ in style from Merian's plates in the earlier caterpillar books, and it is possible that these were etched by Dorothea Maria or another artisan (compare for example Figure 57 to Figure 4).[19] The published text for the 1717 *Rupsen* volume, like that of the two earlier volumes translated from the German, is primarily devoted to very short physical descriptions of the insects and the timing of their metamorphoses. As in the other two *Rupsen* volumes, there is much less content on behavior and little mention of the insects' relationship with various plants, a hallmark of the German volumes. The more compact size of the Dutch and Latin editions meant that all three caterpillar volumes could be bound together, providing another potential cost savings to entice book buyers. It seems likely that the set of three shortened volumes was advertised for sale together.

Later Editions and Their Lasting Effects

Soon after publication of the third *Rupsen* volume in 1717, Dorothea Maria sold all of the copper plates and remaining printed sheets from her mother's insect and flower books to Johannes Oosterwijk in Amsterdam, who published the Latin edition of the *Rupsen* books and a second edition of *Metamorphosis*

---

17   Merian, 1717. *Rupsen.* The third Dutch volume was issued with two earlier *Rupsen* volumes
      and in existing copies, the three are usually bound together.
18   Merian, 1717. *Rupsen.*
19   Other scholars likewise have proposed that Merian had help with the plates for the third
      caterpillar volume; see Reitsma and Ulenberg, *Merian and Daughters*: 231.

FIGURE 57
Merian, 1717. *Rupsen*,
Plate 15. The pupa lacks the
details and more careful
rendering of a similar
pupa in Plate 44 of her
1679 *Raupen* book (see
Figure 4 in Chapter 1). In
addition, the plant here is
more heavily etched than in
earlier plates known to have
been made by Maria Sibylla
Merian.

with 12 additional plates.[20] The plates changed hands again after this, and Dutch and French versions of Merian's European insect books were published in Amsterdam by Jean Frederic Bernard in 1730. In Paris in 1771, Louis Charles Desnos published a French edition that combined the European and Surinamese works of Merian.[21] French edition of Merian's text made her work

---

20   Merian, *Erucarum Ortus* and Merian, Maria Sibylla. 1719. *Dissertatio de generatione et metamorphosibus insectorum surinamensium*. Amsterdam: J. Oosterwijk. Some of the added plates were not by Merian and at least two of these contained misinformation that may have damaged her reputation.

21   The complex publication history of subsequent posthumous editions of Merian's insect books is elucidated by Pick. *Rhetoric of the Author Presentation: The Case*: 95–119. The following editions in Dutch and French combine all of Merian's work on European plants and insects into one book: Merian, Maria Sibylla and Jean Marret. 1730. *De Europische Insecten: naauwkeurig onderzogt, na't leven geschildert, en in print gebragt*. Amsterdam: J. F. Bernard; Merian, Maria Sibylla and Jean Marret. 1730. *Histoire des Insectes de l'Europe, dessinée d'apres nature & expliquée par Marie Sibille Merian*. Amsterdam: Chez Jean Frederic Bernard; and Merian, Maria Sibylla, Pierre-Joseph Buc'hoz, J. Rousset and Jean Marret. 1771. *Histoire générale des Insectes de Surinam et de toute l'Europe*. Paris: Chez L. C. Desnos.

accessible to a much broader audience, as this was the language used by many 18th century scholars. These compilations of Merian's works were printed in a large folio format with four of Merian's smaller plates to a page (Figure 58), and were devoid of any of the preface material of the earlier editions. The French text was based on the shortened Dutch editions, thus missing much of the information on plant/insect interactions included in the original German editions.

Readers of these volumes on the 'History of European Insects' would have been further misled about the nature of Merian's work by the chimeric nature of the volume produced by Jean Frederic Bernard, who combined material from the three *Rupsen* volumes with altered plates from Merian's *Blumen* books. The latter plates were advertised as 'new' material that had not been printed previously, and whether Bernard did not know of the earlier printed *Blumen* books or was simply keen to market his product as something fresh is not known. Further adulterating her work, Bernard contracted with an artisan to add insects to many of the *Blumen* books' plates, and inserted text describing the insects in these flower book images. Most of the awkwardly inserted insects, copied from Goedaert's books on insects,[22] had no relationship to the plants with which they were pictured, thus compromising the integrity of Merian's work. These insertions by Bernard were carried into the 1771 edition as well.

The effect of these corrupted editions of Merian's books was significant and had a bearing on how her work was perceived, possibly lingering until today. Readers with access only to these versions of her writings on insects would have had no knowledge of the more substantial scientific content of her German texts. Moreover, the total number of these later editions in print far exceeded the print runs of the 1679 and 1682 German books, and they were available in more languages.[23] Thus, it is from these posthumous volumes that many scholars such as Carl Linnaeus came to know Merian's work – for better or for worse.[24]

---

22 Van de Plas, Joos. 2019. *Het gestolen kijken – Stolen Observations*. Helvoirt: Published by the author.

23 A count of print editions of the 1679 *Raupen* book catalogued by World Catalog on 11 August 2019 totaled 53. The total for all of the 1730 and 1771 editions was 143. The number of catalogued library copies of the *Rupsen* books and the 1718 Latin edition were not tallied, but the 18th century volumes printed with shortened text were at least four times more numerous than the original German editions.

24 Stearn, William T. 1978. *The Wondrous Transformation of Caterpillars: Fifty Engravings Selected from 'Erucarum Ortus'* (1718). London: Scolar Press: 12.

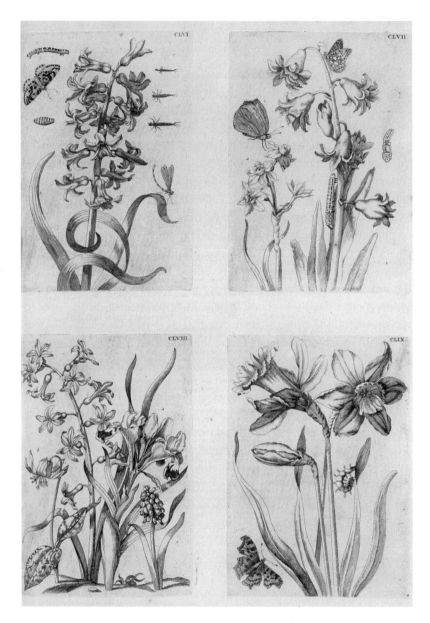

FIGURE 58   Plates 157–160 in Merian and Marret. 1730. *Histoire générale des insectes de l'Europe*. The lower left and upper right plates are unaltered from Merian's original *Blumen* books, but insects were added to the upper left and lower right plates. See Figure 24 in Chapter 1 of this book for Goedaert's images of the moth and caterpillar in the lower right plate.

Recognition and Reception by Near Contemporaries

Merian's reputation as a naturalist grew quickly after her first *Raupen* book was in print. As mentioned earlier, she was recognized for her contributions to art by Joachim Sandrart and to natural history by Christopher Arnold while still living in Nuremberg. A half century later, Nuremberg historian Johann Doppelmayer included her in his volumes on the natural historians of that city. He wrote that after perfecting her painting of flowers and insects "she eventually came thereby to a most extraordinary undertaking: namely, that she paid very close attention, over a long time, to the wondrous changes, the particular plant foods, and more," and that she "made quite new discoveries about such insects for the further advancement of natural science."[25] After the first two caterpillar books were published, she became known beyond Nuremberg, and in December of 1687 while visiting Frankfurt, the German polymath Gottfried Leibniz viewed her 'collection' and her *Raupen* books.[26]

It has been argued that the growing network of naturalists who interacted with Merian as her fame grew was essential to the successful publication of *Metamorphosis*.[27] While not discounting this network as a factor, I would counter that the *Raupen* books were central to her later success with the tropical volume; moreover, that these, along with Goedaert's books, were among the publications that generated fresh interest in insect study and collection. As an established naturalist, Merian had access to collections in Amsterdam such as those of Frederik Ruysch (1638–1731) and Levinus Vincent (1658–1727). In the preface of *Metamorphosis* she cites these collections as having stimulated her interest in travel to Surinam. Merian's network certainly provided her with resources for the production of her Surinam book; Caspar Commelin (1668–1731) native of Amsterdam and botanist at the botanical garden there, provided the Latin names for the plants featured in *Metamorphosis*.[28]

---

25   Doppelmayr, *Historische Nachricht von den Nürnbergischen*: 268–9.
26   Müller, Kurt and Gisela Krönert. 1969. *Leben und Werk von G. W. Leibniz: eine Chronik.* Frankfurt: Vittorio Klostermann: 84. Merian was living with the Labadist community in 1687, but the journal of Leibniz indicates that he may have seen Merian's collection and *Raupen* books at the home of Jacob Marrel's brother (indicated as H. Morel, a variant spelling of the surname).
27   Kinukawa. Natural history as entrepreneurship: 313–27.
28   For more on Merian's connections in Amsterdam, see Van de Roemer, Bert. 2016. Merian's network of collector-naturalists. In Merian, Van Delft, and Mulder, *Merian Metamorphosis*: 19–28.

Interest in natural history and insects in particular was widespread among Early Modern scholars and collectors well beyond Amsterdam, and publication of Merian's books on European and Surinamese insects made her one of the naturalists sought out by visitors to Amsterdam. These visitors included the aforementioned Zacharias Conrad von Uffenbach in 1711 and Jacopo Guidicci, a legate from Cosimo de' Medici in 1714.[29] Her work was sought after by these travelers and others who corresponded with her directly or through intermediaries, and her books and paintings were collected by a range of people, including Russian Tsar Peter the Great (1682–1725), who established the St. Petersburg Kunstkamera and Hans Sloane (Ulster Scots, 1660–1753), whose collection founded the British Museum.[30]

Merian's Influence on Natural History

The effect on a discipline by any given scholar is difficult to trace, but there is evidence that Merian's work had both direct and indirect effects on natural history studies that followed her. She was cited by a number of naturalists, including Linnaeus, who used both her European and Surinam books in his work on systematics, and as mentioned earlier, her books were noted in John Ray's *Historia insectorum*.[31] Many other instances of her influence can be documented, but I will limit discussion to connections with other books on European insects. August Johann Rösel von Rosenhof (1705–1759), another resident of Nuremberg who was trained as an artist, was inspired to publish an illustrated monthly serial titled *Insecten-Belustigungen* after seeing a copy of *Metamorphosis*.[32] In his four subsequent volumes on the natural history of insects, Rösel von Rosenhof cites Merian's *Raupen* books frequently,[33] as does

---

29   Van Veen, Henk Th. and Andrew P. McCormick. 1984. *Tuscany and the Low Countries: an Introduction to the Sources and an Inventory of four Florentine Libraries*. Florence: Centro Di: 44–5.

30   Etheridge, Kay and Florence Pieters. 2015. Maria Sibylla Merian (1647–1717): Pioneering naturalist, artist, and inspiration for Catesby. In *The Curious Mr. Catesby: A 'Truly Ingenious' Naturalist Explores New Worlds*. Edited by E. Charles Nelson and David Elliot. Athens, GA: University of Georgia Press: 47 and 53.

31   Stearn, *The Wondrous Transformation of Caterpillars*: 20. Entomologist Johann Christian Fabricius, a follower of Linnaeus, also referred to Merian's books. For reference to *Historia insectorum* see Chapter 1, note 53.

32   Miall, Louis Compton. 1912. *The Early Naturalists, Their Lives and Work, 1530–1789*. London: Macmillan: 294.

33   Rösel von Rosenhof, August Johann. 1746–1761. *Der Monatlich-Herausgegebenen Insecten-Belustigung*. Four Volumes. Nuremberg: Johann Joseph Fleischmann.

FIGURE 59
Silkworm metamorphosis
on mulberry. Albin, 1720.
History of English Insects.
Plate 12. Albin's silkworm
metamorphosis on mulberry
is very similar in several
aspects to Merian's Plate 1 in
her 1679 *Raupen* book.

another German native, Eleazar Albin (c. 1690–1742), in his volume on moths
and butterflies.[34] Albin also mentions the work of Goedaert, Ray and others,
but his images echo Merian's compositions (Figure 59). Unlike her, Albin focuses more on physical descriptions and less on the metamorphosis or behavior of
his subjects. At least two other German naturalists referenced her work in their
text. Johann Leonhard Frisch (1666–1743) criticized Merian as he did virtually
every other naturalist,[35] whereas Friedrich Christian Lesser (1692–1754) was
favorable in his comments. Lesser cited her *Raupen* books several times, including her work alongside that of Aristotle, Goedaert, René Antoine Ferchault
de Réaumur and others.[36] Merian's invention of pairing insects with their
larval host plants was copied in plates populating many subsequent books,

34 Albin, Eleazar. 1720. *A Natural History of English Insects: Illustrated with a Hundred Copper
    Plates, Curiously Engraven from the Life.* London: Published by the author.
35 Frisch, Johann Leonhard. 1720. *Beschreibung von allerley Insecten in Teutsch-Land : nebst
    nützlichen Anmerckungen und nöthigen Abbildungen von diesem kriechenden und fliegen-
    den inländischen Gewürme.* Theil 1. Berlin: Nicolai.
36 Lesser, Friedrich C. 1738. *Insecto-Theologia.* Frankfurt: Michael Blochberger.

including those of Benjamin Wilkes (English, d. 1749),[37] Moses Harris (English, 1731–1785),[38] John Abbot (American, 1751–1840),[39] and Christiaan Andreas Sepp (Dutch, of German origin, c. 1710–1775). Sepp may not have known the *Raupen* volumes, because he cites the 'History of European Insects' edition in connection with his entry describing a caterpillar that was named for her (*Meriansborstel* in Dutch or 'Merian's brush').[40]

The influence of Merian's books was not limited to those who specialized in lepidopterans, but may be seen in many other natural history studies that followed hers.[41] René Antoine Ferchault de Réaumur (1683–1757) was a prominent French scientist best remembered for his six volume *Mémoires pour servir à l'Histoire des Insectes*.[42] He cites her work, so was familiar with it, and he was the first after her to write in detail of insect behaviors and their ecological relationships with host plants. Like Merian, Réaumur's books showed stages of metamorphosis, specific insect–plant relationships, and insect damage to leaves, however, the images he requested of his illustrators are more diagrammatic than those of Merian.[43] While it cannot be proven that his observations were inspired by Merian's investigations, there is no question that she wrote about environmental factors such as the effects of rain and temperature on the

37   Wilkes, Benjamin. 1749. *English Moths and Butterflies*. London: Benjamin Wilkes. Wilkes cited information from Merian as well as from Goedaert, Albin, Rosenhof and others.
38   Harris, Moses. 1766. *The Aurelian. A Natural History of English Moths and Butterflies, Together with the Plants on Which They Feed. Also a Faithful Account of Their Respective Changes, Their Usual Haunts When in the Winged State, and Their Standard Names as Established by the Society of Aurelians*. London: published by the author and sold by J. Edwards.
39   Abbot, John and James E. Smith. 1797. *The natural history of the rarer lepidopterous insects of Georgia: including their systematic characters, the particulars of their several metamorphoses, and the plants on which they feed. Collected from the observation of Mr. John Abbot, many years resident in that country*. London: Printed by T. Bensley, for J. Edwards.
40   Sepp, Jan Christiaan. 1762–1860. *Beschouwing der Wonderen Gods: in de minstgeachte Schepzelen of Nederlandsche insecten ... naar 't Leven naauwkeurig getekent, in 't Koper gebracht en gekleurd*. Amsterdam: J. C. Sepp. Volume 2: 67–8. This moth (*Calliteara pudibunda*) is featured in Merian's Chapter 47 of her 1679 *Raupen* book.
41   Merian's broader influence on natural history is examined in Etheridge and Pieters, *Merian: Pioneering Naturalist* and in Etheridge, *Merian and the metamorphosis of natural history*.
42   Réaumur, René-Antoine Ferchault de. 1734–1742. *Mémoires pour servir à L'histoire des Insectes*. Six volumes. Paris: De l'Imprimerie Royale.
43   One of Réaumur's illustrators was Hélène Dumoustier, whom he compares to Merian. He expressed great appreciation for Dumoustier's attention to detail and "lifelike portraits" of insects. Terrall, Mary. 2014. *Catching Nature in the Act: Réaumur and the Practice of Natural History in the Eighteenth Century*. University of Chicago Press: 56–8. Terrall's book examines Réaumur's processes, which has parallels with those of Merian.

timing of metamorphosis at least five decades before Réaumur published his
Histoire des Insectes.

Merian's Reputation as a Naturalist

Merian's standing as a naturalist who was cited by others started to ebb in the
19th century, and by the 20th century, she was viewed primarily as an artist
who made beautiful images of plants and insects. As an exception to this, one
early history of entomology lists Merian among a number of other naturalists such as Réaumur as a contributor to early studies of metamorphosis and
insect 'bionomics,' (similar in meaning to ecological studies).[44] Some German
scholars also provide exceptions to this categorization, giving due attention to
her insect studies.[45] Merian's early work on European insects may have been
neglected by modern scholars outside of Germany due to the absence of a
transcription and English translation of the text. Seventeenth century German
printed in blackletter type is not easy reading even for those fluent in modern German. Added to this is the problem of the greatly reduced text in the
Dutch, Latin and French editions of her work on European insects. For three
centuries, these books were more widely available than the German texts that
included her much richer observations on insects and plants. The additions by
various publishers to the posthumous editions further clouded understanding
of her work, and poorly colored copies exacerbated the problem. By the 19th
century, a number of writers actively criticized Merian, and most of these were
referring to posthumous editions of her work.[46] For example, the Reverend
Lansdown Guilding wrote about a 1726 edition of *Metamorphosis*, and his

---

44 Essig, Edward O. 1936. A sketch history of entomology. *Osiris*, 2: 109.
45 See for example Schmidt-Loske, Katharina. 2007. *Die Tierwelt der Maria Sibylla Merian
    (1647–1717): Arten, Beschreibungen und Illustrationen.* Marburg: Basilisken-Presse. An ear-
    lier German scholar includes sections of text from Merian's 1679 *Raupen* book and from
    Metamorphosis in his overview of entomological history, but provides little commentary
    outside of her biography. Bodenheimer, Frederick S. 1928. *Materialien zur Geschichte der
    Entomologie bis Linné.* Berlin: W. Junk: 401–7.
46 Valiant, Sharon. 1993. Maria Sibylla Merian: Recovering an eighteenth-century legend.
    Eighteenth-Century Studies, 26(3): 467–97. Valiant's analysis of Merian's 19th century crit-
    ics and their influence on her reputation is enlightening. She makes a point that 'corrupt'
    later editions of *Metamorphosis* led to criticisms of the accuracy of insect coloration. See
    also Ogilvie, Brian. 2015. Maria Sibylla Merian et la mouche porte-lanterne du Surinam.
    Naissance et disparition d'un fait scientifique. In *Les savoirs-mondes. Mobilités et circula-
    tion des savoirs depuis le Moyen Âge*, ed. Pilar González-Bernaldo and Liliane Hilaire-Peréz.
    Rennes: Presses Universitaires de Rennes: 147–57.

many criticisms included comments characterizing her information and images as "useless," "vile," and "careless."[47]

It is easy to make assumptions about the role that Merian's sex may have played in how she has been considered in studies of the history of science. Certainly, this is a factor that cannot be ignored, but to assign sexism as the sole reason that her scientific contributions are not well known today would be to oversimplify a number of complexities. The combination of the truncated text of her observations in widespread editions of her work and the derogatory comments of some 19th century writers had lasting effects on how her work was perceived. Another factor that may have led her work to be undervalued by early science historians could be the focus of her research. Merian's attention to the relationship of insects and their plant hosts and her disinclination to contribute to classification schemes, was out of the mainstream in Early Modern natural history studies. But whatever the reasons, science historians have paid scant attention to her research on insects, and she has received little credit for her contributions to understanding plant/insect interactions. Hopefully, closer reading of the original 1679 *Raupen* book and further examination of her information-packed images will increase appreciation for the quantity and quality of ecological and behavioral observations that she made.

---

47   Guilding, Lansdown. 1834. Observations on the work of Maria Sibylla Merian on the insects etc. of Surinam, *Magazine of Natural History and Journal of Zoology, Botany, Mineralogy, Geology and Meteorology*, 7: 355–75. It should be noted that Guilding questioned Merian's accuracy even though he never set foot in Surinam.

PART 2

Plates, Translation and Commentary

∴

FIGURE 60   Illustrated title page of Merian's 1679 *Raupen* book

Maria Sibylla Merian's Caterpillar Book

Introduction

The core of Part 2 of the present volume is an annotated translation of Maria Sibylla Merian's 1679 *Raupen* book, presented with each of her fifty chapters adjacent to the accompanying image. The order of Merian's materials in each bound volume could of course vary, because printed books in her time typically were sold as uncut sheets, which the buyer would have bound after purchase. However, Merian undoubtedly meant for each chapter of her text on insects to be interwoven with the plates illustrating her findings. In this way, her readers could use the descriptions to understand the images and vice versa. The order in which we present her text and images follows that of most copies of the book that I have seen with one slight change. We chose to place the two poems by Christoph Arnold together at the beginning of the material in order to simplify discussion of these. In most existing copies of the 1679 *Raupen* book, Arnold's 'Caterpillar Hymn' is bound into the book just before Merian's indices.

My commentary is inserted after each short chapter of Merian's text and its corresponding plate. The discussion in each entry is meant to highlight her novel natural history observations, but also to address her work in a historical context. Various images or sections of her text are used to illustrate a particular practice by Merian or as an example of some type of observation (e.g. on behavior, metamorphosis, or the effects of weather). In this way, both her specific and the general contributions to natural history can be elucidated and evaluated. The commentary also addresses her way of organizing information and emphasizing certain aspects that she found important, intriguing, or amusing.

Merian's Plates

The plates from Merian's 1679 *Raupen* book are represented here by beautifully hand-painted counterproof copies of her plates. This collection is held in the Library of the Netherlands Entomological Society at Naturalis Biodiversity Center in Leiden, and they are reproduced here by courtesy of the Netherlands Entomological Society.[1] Most of Merian's books were sold with uncolored

---

1 Merian, 1679. *Raupen*. The Library of the Netherlands Entomological Society at Naturalis Biodiversity Center is abbreviated in image credits for the plates as NEV Library/Naturalis.

plates, and digitized copies of these black and white plates can be found in a number of places online. Because images of uncolored plates are readily available for viewing, we chose these colored counterproofs to illustrate her caterpillar book. In these plates, the insects are painted with colors very similar to those in her study journal paintings. Thus, the insects are represented largely as they appear in nature and as Merian observed them. Other colored copies of Merian's caterpillar books exist, but most were printed editions that were painted post-purchase by an unknown hand. In many existing copies, this was done poorly, and the insects and plants look nothing like those of their counterparts in nature (e.g. light brown wing color is sometimes represented as pink or blue).

The counterproof images that follow are mirror images of Merian's printed plates. She did not refer to left and right in her text, and so the reversed counterproof images do not pose a problem for the reader. Someone added a black line border to these particular counterproofs that is not present in the usual printed plate. The illustrated title page in the Leiden edition is not a counterproof of the original, but is instead a watercolor painting that differs substantially from the printed version of this page. I chose to omit the painted title page and instead use an uncolored print from a different volume (Figure 60). In order to approach the appearance of Merian's original book, the images herein bear her plate numbering system and the captions for these illustrations are abbreviated. In the printed plates, Roman numerals designating each specific plate appear in the upper right corner. To avoid reversing the plate numbers, these were masked for printing the counterproof images, and someone hand-labeled the plates. The modern names of the plant and animal species are included in the commentary sections rather than with the plates or with her translated text.

Notes on the Translation

On the title page of her 1679 caterpillar book, Merian addresses an audience of "naturalists, artists, and garden lovers." At various places in the text, she offers observations or suggestions intended as "useful information for attentive

---

The counterproof copy of the 1679 volume is one of a set of three hand-painted volumes owned by Rotterdam merchant Adriaan Wor in 1774. The 1683 volume is also a counterproof, but the third volume is instead a painted copy of the third *Rupsen* volume (1717). Oddly, the Dutch text with each volume is a calligraphic copy from the truncated *Rupsen* editions. For more on these unusual volumes, see Reitsma and Ulenberg, *Merian and Daughters*, 167.

gardeners" (Chapter 18). From time to time, she injects mention of some popular belief or cultural practice (e.g. Chapters 20 and 34) or a personal reaction of disappointment or delight in her work. Her German – neither academic, ornate, nor obscure – is conversational in tone and clear enough for the modern reader of the language. In translating it into English, we have strived for accuracy and clarity, but also for fidelity to the spirit of her presentation.

Merian's book title gives important indications not only of what she observed and studied, but also of how she worked. Some key words in the original, seemingly clear in our translation, could be imprecise or ambiguous for modern readers of German. A familiar English word, 'invention,' (*Erfindung*) appears on the title page, but without any hint of a device, technique, or process to which it might refer. In 17th century German, however, the meaning of *Erfindung* was less well defined, and at times used interchangeably with *Entdeckung* (discovery). We take Merian's "new invention" to mean her original ecological compositions combining insects and their host plants; therefore, our translation recognizes the form of her presentation as an invention, not a discovery.

Of course, the language of the biological sciences systematized over the more than three centuries since Merian's time has features not found in the German or English of the Early Modern period, particularly in the terminology for organisms, their structures and processes. Thus, we have tried to use only English terms that we can verify as having been in general use by the late 17th century. Some are Merian's preferred substitutes for objects and events: 'date seed' for pupa, 'summer-bird' for butterfly, (both of them names she would have heard, not made up herself); 'transformation' for metamorphosis; and 'change' for pupation. These terms we have kept as she defined them and for the reasons she gives for using them (see Preface and Chapter 14).

In some instances, she uses terms that readers of her book today would find unusual and perhaps confusing. She introduces several of these in her preface, an indication that her contemporary audience may have needed clarification of them as well. For example, the German word *Vöglein* or *Vögelein* is generally understood as 'little bird,' but it may also refer to a winged insect, especially a butterfly or moth.[2] When used in reference to the adult stage, *Vög(e)lein* may be translated as 'adult' or 'insect,' and we have used both of these (historically inappropriate) terms in the interest of clarity for the modern

---

2   The word does occur in English, e.g., in the case of coccinellids, called ladybirds in Britain and other parts of the Anglophone world.

reader.[3] 'Summer-bird' is the translation of *Sommer-vögelein*, a butterfly, "since for the most part they fly in summer."[4] 'Moth-bird,' from the parallel form *Motten-Vögelein*, is the name for a moth, although she occasionally uses *Motte* alone.

Merian has two different terms for insect eggs. The usual German word, *Ey* (pl. *Eyer*, dim. *Eylein*), appears with that meaning in the chapters that follow, but in her preface she uses it only in reference to the cocoons and 'capsules' of the pupal stage. A second term, *Samen*, means 'seed,' but in the preface and in several of the following chapters she employs it, in the singular form, particularly in reference to the clutch of eggs laid by a female insect (e.g. Chapters 1, 5, 45). In such instances, we have translated in accordance with her context to avoid ambiguity.

'Date seed' for the German *Dattelkern*, used in referring to a pupa, is a literal translation that we have kept throughout the text. The modern German name, *Puppe*, seems not to have been in currency at the time, but *Dattelkern* clearly was, as Merian explains in the preface: "I have ... chosen to retain this word because the greater part do resemble date seeds, and because I have heard them generally called that since my youth."[5]

Certain processes or actions described in the German text have more than a single name, typically a word pair, and we could not always be certain of the distinctions between them. Three such pairs that occur profusely in the work are: *Veränderung / Verwandlung* (change / transformation); *zeichnen / malen* (draw / paint); *radieren / stechen* (etch / engrave). At times it appeared that Merian used *Veränderung* to denote just the pupal stage in the life cycle, while *Verwand(e)lung* referred to the complete process, as in her title, *Der Raupen wunderbare Verwandelung*. But that distinction did not hold in every case, and it became a matter of context-based choices for the translation. The verbs *zeichnen*, *malen*, *radieren*, and *stechen*, techniques in the production of Merian's images, are more difficult to sort out and depend as much on what we know of her working methods as on lexical equivalents.

A pervasive feature of Merian's text that is seldom evident in the translation is her frequent use of the diminutive form of nouns. A prime example is *Vögelein*, but there are many others formed with the characteristic southern German suffix, *-lein*. The tiny organisms and parts of organisms she is describing rarely require qualification as small in English. She occasionally uses both

---

3   Nor does Merian have the word *Insekten* available to denote the taxonomic class of animals; for that, her usual term is *Thierlein* – "little creature(s)."
4   Merian, 1679. *Raupen*, preface: p. iii.
5   Ibid.: p. iii.

forms to designate more than one size or variety of insect, as in Chapter 25: *unzählbare Raupen und Räuplein* (countless caterpillars, large and small). In such cases we have honored her obvious intention. Parenthetical numbers in the translation of Merian's text refer to the page numbering in the original.[6]

The
Wondrous
Transformation
and Particular
Food Plants
of
Caterpillars,

Wherein,
by means of an entirely new invention
the origin, food, and changes
of caterpillars, worms, butterflies, moths,
flies, and other such small creatures,
together with their time, location, and characteristics,

For the benefit of naturalists, artists, and garden lovers,
are carefully investigated, briefly described,
painted from life, engraved in copper,
and personally published
by

Maria Sibylla Graff,
Daughter of the late Matthaeus Merian the Elder

To be had in Nuremberg
from Johann Andreas Graff, Painter
in Frankfurt and Leipzig from David Funk
Printed by Andreas Knortz, 1679

---

6 Our translation is based on the copy of *Der Raupen wunderbare Verwandelung* held by the Heidelberg University Library. https://digi.ub.uni-heidelberg.de/diglit/merian1679bd1. Accessed 26 February 2020.

Song of Praise

It is remarkable that women, too, would venture
　　to treat the very matters
　　　　with serious intent
　　that scores of learned men have pondered without end.

What **Gessner, Wotton, Penn**,[7] and **Moffet** did not publish
　　for us to read today;
　　　　is brought to light, O **England**,
　　by a gifted woman's hand, in my own **Germany**.

What **Goedaert** and **de Mey** in **Zeeland** once did write,
　　we read with grateful pleasure;
　　　　But praise is due no less
　　to a woman who aspires to do the same as they.

What **Swammerdam** promises, what **Harvey** once did lose,
　　is now made known to all;
　　　　for an ingenious woman
　　has done all this herself, as if in idle hours.

What the **Italian, Redi**, has lately come to know;
　　and **Florentine Stradano**[8]
　　　　so many years ago
　　engraved in metal there; for her is but a trifle.

Though **Spain** may well praise **Bustamante**[9] highly;
　　we'll test by equal standards,
　　　　and show what can be done
　　by this industrious **daughter** of worthy **Merian**!

C. ARNOLD

7　Thomas Penny (1532–1589). Arnold, in the original, writes names as: Gesner, Wotton, Penn, Mufet, Gudart, von Mey, Swammerdam, Harvey, Redi, Stradan, and Bustamantin (the last with inflected accusative-case ending – names of persons in this period were generally subject to the same grammatical inflections that applied to German common nouns).

8　Jan van der Straat (Bruges 1523–Florence 1605).

9　Juan Bustamante de la Cámara, Spanish naturalist, fl. 16th c.

Caterpillar Hymn

To the melody:
Jesu, der du meine Seele[10]

1.

Lord, of every thing Creator,
　　all the wonders Thou hast wrought
in Thy wisdom, I will sing them,
　　works that beggar human thought.
These are all beyond comparing,
and surpass my understanding.
　　Peerless in unnumbered ways,
　　Thou alone receive the praise!

2.

This is my heart's joy and pleasure:
　　o'er the broad expanse of fields
feasting eyes in fullest measure;
　　and the world in fairest flow'rs,
gracing every bush and meadow,
thrills my sight for endless hours:
　　Blossoms everywhere I gaze;
　　myriad creatures, boundless praise!

3.

Shall all these laud their Creator,
　　and I utter not a word?
If the pot extols the potter,
　　I confess then: Thou art He!
Shall I sleep and fail to greet Him
who created me and all men?
　　Shall I not my praise outpour,
　　and be silent? Nevermore!

10　Chorale text and perhaps melody (1641) by the early Baroque poet and pastor Johann Rist
　　(1607–1667). Rist was a member of the *Pegnesischer Blumenorden* (Order of the Flowers of
　　the Pegnitz [River]), the Nuremberg literary society to which Christoph Arnold belonged.

4.

Thus my heart is set to singing!
  Brightly colored little birds,
dressed in splendor by their Maker,
    shall bear witness to my vow:
to respect His every creature,
and observe with keen devotion
    every thing that lives on earth,
    where GOD's goodness sheds its light.

5.

Just behold the modest flowers
  He gives to the worms for food;
all the blossoms, leafy branches
  Nature offers to them here
must sustain so many thousands,
teaching us GOD's caring mercy:
  See the flow'ring meadowlands
  clothed in regal splendor now.

6.

Gold and silver I see shining,
  pearls adorn their flowing gown;
lovelier than we could view them
  painted by a master's hand;
softest velvet, silken cover
clothe their every wave and furrow.
  And our sense cannot discern
  each that spins and mates in turn.

7.

Dearest GOD, Thou wilt reward us,
  every one, in his own time;
like the caterpillar changing,
  and by its mortality,
once again to life returning
like the dead from earthbound slumber:
  Grant that I, poor worm, may be
  raised again to life in Thee!

C. ARNOLD

Front Matter and Indices

The illustrated title page in the first edition of Merian's 1679 *Raupen* book includes the short version of the book's title centered in a wreath of white mulberry. Two short branches form the wreath, and the overlapping ends of the
stems at the bottom of the plate are inscribed with both her married and her
maiden name. Maria Sibylla has been shortened to 'Mar. Sibÿll' with the remainder of her given name disappearing under part of the branch. Her married
name is extended to Gräffin, the feminine suffix *-in* having been commonly in
use at the time. Her family name is likewise inscribed as Merianin, and it is
cleverly placed on the upper stem end such that it also reads as part of her full
name. The inscription "*geb.* Merianin" (*née* Merian) is at least as prominent as
her married name, and this was likely intentional, as the book's title page also
emphasizes her relationship to the eminent Merian family firm of engravers
and publishers headed by her father. The significance of the insects chosen
for inclusion in this plate is not clear, and indeed, they are more generalized
in appearance than the insects within Merian's numbered plates. In addition
to the lepidopterans, she includes a fly and a small beetle, neither of which is
identifiable. The caterpillars have been thought to be silkworms by some, and
this would make sense given the mulberry wreath, but the two different types
of adults shown at the top are not clear representations of silkworm moths,
and neither they nor the caterpillars can be identified. None of these nonde-
script caterpillars are actively feeding on the mulberry, although she makes it
clear that the mulberry has been attacked by insects, showing several leaves
with large holes. Rather than featuring specific types of insects in this title page
image, Merian indicates that her book contains a variety of insect types. She
makes it very clear that insects hatch from eggs, showing this in two places on
the mulberry wreath.

    The printed title page includes the book's full title. If the order of phrasing is
meant to convey emphasis, then Merian wants her reader to understand that
her focus in the book will be the transformation (metamorphosis) of caterpillars and the importance of their food plants. Her "new invention" can be taken
to mean her original compositions combining insects and their host plants.
The title continues by delineating some other types of creatures and information included in the book, and the wording stresses Merian's personal contribution through her investigations of nature, her art, and her publication of
the book. Invoking her well-known father's name lends further weight to the
project. Johann Andreas Graff is identified in this section of the title page as a
painter, and the primary person from whom the book may be had. David Funck
is listed as a second vendor of the book in Frankfurt and Leipzig, broadening

the range of its availability.[11] The printer, Andreas Knortz, and date of publication are acknowledged last.

Christoph Arnold wrote two celebratory pieces that were published within the *Raupen* book. In the six short stanzas of the first poem, Arnold favorably compares Merian to eleven men who studied insects; among these, he includes naturalists Conrad Gessner, Johannes Goedaert, Johannes Swammerdam and Francesco Redi. In the last phrase, he also links Maria Sibylla to her well-known father, Matthaeus Merian. Arnold's other contribution to the 1679 *Raupen* book was a seven-verse poem or 'hymn' praising God's creation of nature, and the beauty and wisdom exhibited therein. Caterpillars are extolled in the title of the poem and in the verses, the beauty of flowers is celebrated, and the 'Caterpillar Hymn' ends by using the emergence of a butterfly from slumbering pupa as an allegory for resurrection, much as Goedaert had done.[12] Arnold's praises of Merian continued in a third poem composed for her 1683 *Raupen* book.[13]

Merian's book has two indices, which are included at the end of the translated text. The first index is something of a catch-all, which she terms "the most noteworthy items in caterpillars' transformation and food plants." Organized alphabetically by the term of interest, this index consists primarily of plant and insect common names and the colors and physical appearances of the insects. However, Merian includes other terms that she considers important, including lightning and sugar. The second index, titled *Arborum, Fruticum, Plantarum, Florum and Fructuum*, is more straightforward and consists of the alphabetized list of the Latin names of plants included in the book. Merian must have had a hand in generating the first list, choosing the information that she wished to index, and the printer could have added the page numbers after the pages of type were laid out.[14] However, the index of plant names may have been prepared by the printer working on his own and using the Latin phrases

---

11   David Funck lived and worked in Nuremberg, but he may have had bookselling connections in the other cities. Grieb, Manfred H. (ed). 2007. *Nürnberger Künstlerlexikon: Bildende Künstler, Kunsthandwerker, Gelehrte, Sammler, Kulturschaffende und Mäzene vom 12. bis zur Mitte des 20. Jahrhunderts.* Munich: K. G. Saur: 439f.

12   The list of naturalists in the "Song of Praise" corresponds with works found in Arnold's library, and gives us a very small sample of the wide variety of books to which Merian would have had access as a close acquaintance of Arnold. Arnold's books on nature and medicine are catalogued in Jürgensen, *Bibliotheca Norica*, Volume 1: 779–98.

13   Merian, 1683. *Raupen.*

14   In the Latin index included in this book, the page numbers correspond to those of Merian's 1679 text. Two of these have been corrected from errors in the original index, which listed p. 37 for the mulberry plant appearing on p. 1 and p. 42 for the sweet cherry on p. 47. Plant name order has been changed from the original in a few instances to maintain

heading the right hand page of each of her chapters. Throughout her book, Merian often indicated whether a fruit-bearing plant was shown as flowering or fruiting, and these designations were carried into the Latin index. Only Plate 17 does not bear the name of a plant, but instead Merian's appellation for the lappet moth caterpillars, *Vermes miraculosi*. Merian did not identify the food source for her "miracle worms," and so did not include a plant image or name. The clue that the printer may have prepared the index of Latin plant names comes from the inclusion of *Vermes miraculosi* in that listing; the printer likely had no way of knowing this was not consistent with all of the other entries in the second index, which were plant names. On the verso of the final page of all copies of the 1679 edition that I have examined is a list of errata.[15]

consistency with her chapter headings (e.g. *Hyacinthus orientalis caeruleus* rather than *Hyacinthus caeruleus orientalis*) but alphabetical order is unaffected.

15 Merian, 1679. *Raupen*: unnumbered page. "*The bookbinder should arrange the copperplates according to the capitalized numerals of the chapters; and the reader should correct several errors as follow: Sheet 4* [of the preface], *Line 2, read: über den. 12, 6: aber eine. 14, last line: etliche Eylein. 16, 7: sechs braune. 27, 2: pruni. 28, 15: Tag. 53, 25: Nachdem. 68, 23: kreucht. 68, 24: Eylein. 92, 20: wieder.*" All of these corrections have been incorporated in the translation of Merian's text in this book.

Most Esteemed Reader and Lover of the Arts

While I have always endeavored to adorn my flower paintings with caterpillars, summer-birds, and small creatures of that kind, as the landscape painters do with their pictures, to make the one, as it were, more lifelike by means of the other, I have also often taken the trouble to collect them; until eventually, by observing the silkworm, I became aware of the caterpillars' changes and began thinking about them and whether that very same kind of transformation might not occur in others as well. Since then, after diligent and painstaking investigation, I concluded that their manner and type of change is nearly identical; the difference is that silkworms spin useful filaments while the others make useless ones; and indeed many caterpillars or worms even turn into flies or gnats, which to my knowledge does not occur with the silkworms. Thus I continued to observe them into the fifth year and discovered astonishing changes, keeping them all in boxes and showing them to anyone wishing to see them. Now every time I did so, they acknowledged and praised most highly God's great power and marvelous care for such insignificant creatures and useless winged insects. All of which then finally so moved me and caused me—especially having often been urged and encouraged by learned and respected persons—to present such Divine marvels to the world in a modest book. Therefore, seek herein not mine, but God's glory alone, and glorify Him as Creator of even these smallest and least of worms, for they have their origin not in themselves, but from God, who endowed them with such wisdom that in some things (as it would seem) they almost put humankind to shame: by faithful adherence to the time and sequence of their lives and not emerging before they know how to find their own nourishment. Thus the insects almost never lay their eggs (2) somewhere other than where they know that their young will find their nourishment, or food.

Not that it has not cost me great effort and time to search for such small creatures and give them their food for many days, even months. For if they do not receive their accustomed nourishment they either die or spin themselves up [in cocoons]. I have had to draw some of them right away, some later when already half changed, and others fully and completely changed; and paint them again as soon as they had spun [a cocoon] or turned completely into a hanging or lying date seed, and then wait to see what would finally come of them. And if anything irregular occurred, I have not allowed myself to be discouraged by the utmost care and effort of illustrating it once more from life; and if false changes occurred in the process, then faithfully drawing those as well. And I further undertook, with the generous help of my good husband, to include paintings from nature of the food for each type.

Even though everything I set out here can also be found in the descriptions which follow, I consider it necessary to express much in few words: in general, all caterpillars, as long as the adult insects have mated beforehand, emerge from their eggs, which have the appearance of millet seeds; and the young caterpillars are at first so small that one can hardly see them. But their size increases by the day, especially when they have enough food. Some then attain their full size in several weeks; others can require up to two months. Many push off their skin completely three or four times, just as a person removes a shirt over his head, and change to different colors; some reappear as they were before. The maggots or worms, on the other hand, have their origin for the most part in decayed caterpillars or other filth, and also in the caterpillars' droppings. They do not stay that way for long, but soon spin and change into brown capsules, from which all sorts of flies will later appear. Of the caterpillars, however, we can actually note that some (3) spin themselves up like a silkworm and produce exactly the same kind of capsules encased in silk, although not as strong as that of the silkworm. Some, however, undergo a very different change, in that they attach themselves, hanging head down, to a wall, leaf, stem, or other fixed place of that kind; and, having stripped off their skin over their heads, form themselves in the space of just a few hours to look like a swaddled child, so that one can plainly see. Some appear as if they had been clad or sprayed with gold, whence some people rightly call them *goldlings*; and some appear like silver or mother of pearl, so that one could well refer to them as *silverlings*. Some lay themselves down on a leaf or some other surface, having also pushed off their husk, but without first wrapping the capsule and without attaching themselves so as to hang, and then look almost exactly like a date seed. I have (may the kind reader take note) chosen to retain this word, because the greater part do resemble date seeds and because I have heard them generally called that since my youth. However, if there is a spun cocoon lying nearby, I will use cocoon (as indeed it is and so appears) or capsule to designate it. I will also keep the term capsule for maggots or blowfly-worms.

It is useful to know that a hanging date seed is attached by a thread so strong that it almost requires a knife to cut it off, although doing so will hinder its maturation. They now remain for some twelve to fourteen days, some even the entire winter, until cold temperatures are past and they feel the warmth of the sun. The warmth then matures them quickly so that summer-birds come forth from the date seeds. Others call them *Butter-Vögelein, Zwifalter, Fledermäuse*

[bats], or such; but I will refer to them as summer-birds, since for the most part they fly in summer and that name is thus appropriate for them. But the adults from those caterpillars that change into a cocoon or date seed lying flat on some surface, I will refer to (4) as moth-birds, to distinguish them from the former. These are nearly immobile during the day, flying about mostly at night, and can be recognized by their large heads. The rest I will refer to as gnats or flies.

Now in all of these I have noted that they do not require more than an hour to grow to their full size, and that when they come out, their wings are very small and soft; but we can see them attain their full size and strength in a half-hour to an hour, so that they are able to fly where they will. The moths in particular at this point, like the silkworm moths, are remarkably ready to mate, something no caterpillar or worm does, going only in search of food. In all of this I have never been able to find caterpillars or worms that had eyes or could see, while the adult insects have very bright eyes. Moreover, I have also observed that they can be kept fairly well with sugar, extracting it or other sweet substances with the long beaks located prominently between their eyes. After their initial growth they become neither larger nor smaller. Their dust, which is their color, can be wiped off, even somewhat damaged, by themselves, a wing broken or bent. As for their location, summer-birds and moth-birds prefer to stay near the plants, flowers, and fruits that serve as their food, so that they can soon lay their eggs on them again.

Finally then, the following fifty copperplates are here presented as best I can engrave the living things in black on white. But if the nature- and art-loving reader wishes to have the entire plates, or only the caterpillars and their changes, along with the insects, accurately colored, we will gladly supply either one. And so,

> The beginning has been made, and if it is your wish,
> I will pursue this work in service to my readers,
>   that I may keep their interest keen with art
>   and earn the praise and favor of great men.

Preface

In the preface of her first caterpillar book, Merian makes the case for her study
of nature, both from the objective of honoring God's power and care of even
the "insignificant and useless winged creatures" and with the goal of creating a
work that would serve her readers. For this dual purpose, she wished to bring
"such Divine marvels to the world in a modest book." Her readers are addressed
initially as lovers of the arts and at the end of the preface as both nature- and
art-loving. The preface closes on a telling note, her plan to pursue further the
work of studying insects, both for her readers and to "earn the praise and favor
of great men." Hints of this ambition peek through in other places in her writing, scattered among her more pious and humble statements, where for example, she leaves some unexplained phenomenon to "learned men." Merian
establishes her credentials as a dedicated and careful observer of nature by
delineating in the preface and elsewhere the laborious and time-consuming
nature of her work. However, she cautiously tempers this statement with an
expression of thanks for the help of her husband, a personal acknowledgment that does not reappear in her later books. Some have taken mention of
her husband's help within the sentence stating that her book included plants
painted from nature to mean that Graff contributed paintings of plants. This
conjecture seems unlikely for a number of reasons. First, no work by Graff in
various collections includes such botanical images; he was known for his city
scenes (see Figures 38 and 39) and landscapes. In the latter, plants like trees
appear but they are not particularly naturalistic or well observed. More telling
is the fact that Merian's images of plants after her separation from Graff look
much the same as the ones in the 1679 *Raupen* book. Perhaps her thanks to him
in this context was for plants that he brought home for her to paint; alternatively, it simply could be awkward phrasing as occurs from time to time in her
writing.

Merian uses the preface to establish baseline biological information for
the chapters that follow. She writes that she began by observing the silkworm,
which led her to wonder if other caterpillars would undergo similar changes.
Her "diligent and painstaking investigation" led her to conclude that with the
exception of some anomalies, this was indeed the case. After describing her
working methods, she goes on to explain the life cycle of what we now call
lepidopterans. She begins by asserting that caterpillars emerge from eggs laid
by adults who have mated beforehand. In a time when insects were generally
believed to be produced by spontaneous generation, this was a bold statement

with which to begin. She then tells readers how the tiny caterpillars grow when their food is available, and how they molt their exoskeleton between larval stages, describing this process as "pushing off" their skin. Merian is careful to explain that within this general pattern there is variability in both length of time and number of molts by different types of caterpillars, and that the caterpillars may change color through this process. The pupal stage is then described, introducing the reader to the variation in cocoons and pupal casings and to the different locations of pupation, be it within a rolled leaf or attached to a surface. In her 1705 book on tropical insects,[1] Merian uses the word pupa for this developmental stage. However, in this much earlier book she explains that she will use the descriptive term "date seed" for this structure, a holdover from her youth, and in her eyes, an apt match for the pupal appearance.

As mentioned earlier, there are alterations in the typeface that bring attention to structures like date seed and cocoon, and in the preface, these terms are in bolder font. Merian does not use the term metamorphosis in this work as she does in the title of her 1705 book, and in describing the overarching process tends to use 'transformation' (*Verwandlung*). She uses the term 'change' (*Veränderung*) as well, and there is no clear delineation between how she uses the two terms, although the latter appears a bit more often in the context of pupation. After the onset of the caterpillar's pupation, Merian patiently waited to see what type of adult moth or butterfly would emerge, hoping to complete her cycle of images and text for each type of caterpillar. On several occasions however, an unexpected kind of insect would appear from a caterpillar or pupa, rather than a moth or butterfly. In her preface, Merian makes the point that she faithfully depicted these false or 'disorderly' changes, one of many ways in which she testifies to her objectivity. She goes on to address the emergence of the "maggots or worms" from some caterpillars and the fact that these worms do not go through the same kind of development as do larvae that eventually become moths or butterflies. Even as she was perplexed by the disruptive and unpredictable appearance of what we now understand as parasitic invaders, she duly noted their differences from her primary subjects and specifies for her readers that she would designate their pupae as 'capsules' rather than as date seeds. By the time Merian's third caterpillar book was published in 1717, she understood that these organisms were flies and wasps that had parasitized the lepidopteran larvae or pupae, laying their own eggs in the host.

---

1   Merian, *Metamorphosis*.

Merian studied her subjects in gardens and meadows as well as raising them in her house. The preface introduces her observations on the effects of season and temperature on insects, and the variable ways in which different lepidopterans respond. Some adults emerge in summer after days or weeks of pupation, and others overwinter as eggs, caterpillars or pupae. In speaking of these pupae, Merian wrote that this stage lasts until "cold temperatures are past and they feel the warmth of the sun." Such observations speak of the careful and time-consuming husbandry needed to record such critical life history information. From her discussion of temperature effects Merian moves on to establish a dichotomy between what we now categorize as butterflies and moths, but that she calls summer-birds and moth-birds respectively. She elaborates on the differences between these groups in later chapters, but here notes that moth-birds are inactive during the day and fly at night in contrast to the day-active butterflies. She also mentions that moths have larger heads than butterflies, and observes that moths tend to pupate on flat surfaces.

She concludes her introductory description of metamorphosis with a brief step-by-step account of the emergence of adults from their pupae. Here again she distinguishes between butterflies and moths, remarking that the latter often emerge "remarkably ready to mate." She then emphasizes the fact that caterpillars never mate, but instead go off in search of food after hatching from their eggs. In the chapters that follow, Merian frequently remarks on the ravenous nature of the larvae, and lists the plants on which they feed, something she does not do for the adults. She notes that adult moths and butterflies do not grow in size after emergence, something that is in clear contrast to caterpillars, which grow constantly until pupation. In two places in her preface Merian states that the adult insects lay their eggs on food plants, and in the first mention, attributes this to God-given wisdom that results in their young having food as soon as they hatch from their eggs. She is correct insofar as it is generally true that lepidopterans tend to lay eggs on the larval host plants; however, the food plants of the adults often are different from those of the caterpillars. Merian focused primarily on the larval plant–insect relationship in her studies, because establishing the feeding habits of the many different caterpillars was critical to her success in obtaining the complete life cycle of each species.

Despite beginning and ending the preface with references to art, Merian sets the tone for an empirical work consisting of first-hand observations and 'faithful' depictions, and she makes it clear that the text and plates are the result of her own labors. The expression of spiritual beliefs and appreciation for the assistance by her husband make up only a small portion of this introduction, and Merian's clear passion for the work tempers somewhat her assertion that she wrote the book due to encouragement by others. As the chapters unfold, her intense dedication to the study of insects is increasingly revealed, and it becomes clear that Maria Sibylla was motivated by the same innate curiosity that drove naturalists before and after her.

PLATE 1   Merian, 1679. *Raupen*

1. Mulberry tree with fruit *Morus cum fructu*

(1) Since nearly everyone knows that the silkworm is the most useful of all the caterpillars and, as the noblest, far surpasses them, I have chosen to begin the present work with it and to make the title page of this book in honor of it. In the wreath on the title page, the red mulberry is shown with only young leaves together with its blossoms,[1] on which can be seen the eggs [*Samen*] and several newly-hatched caterpillars, which cannot feed on leaves any larger at that time of year. In this first copperplate, however, I have shown the white mulberry with its green leaf, on which rests a large silkworm ready to enter its next change. Its color, ordinarily white, is becoming yellowish, shrunken, and somewhat transparent because it is preparing to spin [its cocoon]. It begins swinging its head back and forth and releasing the silk threads from its mouth. At that time it is placed in a little paper house which is pointed at the bottom but wide at the top, like a *Scharmützel* [cone], so that the work is made somewhat easier.[2] People who keep large numbers of them will place small branches from trees (on which they shall spin) in a special room and lay the leaves over that, so that their droppings can fall to the ground. Now when the time comes and a caterpillar finds no suitable place to spin, (2) or is overfed, then it does not spin but shrivels and turns into a date seed without a cocoon. But otherwise it spins very busily and remains hard at work until it completes its entire cocoon, which is oblong and either white, yellow, or greenish. If it is white, then its round legs will have been that same color; but if it is yellow, they will have been yellow; and if greenish-yellow, that is how they will have appeared. After completion of this [cocoon], it becomes a date seed in order to change into a new moth-bird which, when it has come forth, is usually white in color and needs half a day to get clear, dry wings, or its finished form. It has only six legs, two brown horns, two small brown eyes, and four white wings in which brownish stripes can be seen. The male is more delicate and smaller than the female, which has a fatter body, the male a thinner one. As soon as they have attained their full strength they mate, and that same day or the day

---

1 Although Merian here uses the German word *Blüe* (*Blüte*: blossom, or season of flowering), the title page in fact shows immature fruit.

2 Both the text here and the corresponding section from her study journal (Appendix, entry 1) use the word *Scharmützel* for the paper cone, and in the journal she notes that the term is a local one from the Nuremberg region. The correct form, however, is *Scharnützel*. Since this apparent error in spelling occurs in both sources, we might conjecture that Merian, a Frankfurt native and unfamiliar with the word when she moved to Nuremberg, misheard "m" for "n" and wrote it that way.

after and for several days following, they lay little round, yellow eggs like grains of millet, and the adults then die. On each egg there is a tiny dot of brownish color. One can also soon tell whether something useful will come forth from them; nothing will come from any that are collapsed or crushed, or appear like empty husks, for they are spoiled. Now if one does not want the new (3) silkworms to eat, they should be kept in a cold place, but otherwise in a warm one; or they can be placed in the sun, or even in warm rooms. Then the little worms, which chew their way out when ready, crawl out of the aforementioned dots. But the time (in which month) this occurs is determined as described above. Here in our region they come forth sometimes in April, sometimes in May, according to the warmth received early or late. People whose worms emerge prematurely feed them salad [lettuce] right away;[3] and in moving and cleaning to give them fresh food (since the time must then be carefully observed), they treat them very gently, because they are so delicate. Some do this using small pigeon feathers, some with small brushes. But they must also not be given any wet leaves, for as soon as they eat something spoiled or wet they become sick and die; thus, leaves that have been rained on must be carefully dried off. When the little worms are barely four days old they molt, and many die in the process. Also, when the time comes that they are started on mulberry leaves, they molt a second time. And the number of times this occurs can vary. If there is a thunderstorm and it begins to lightning, they must be covered; otherwise they get a disease which some call jaundice, because they begin to turn yellowish and become lethargic in feeding and finally die. It is also better not to mix the leaves of red and white mulberries for them, for it is said that they will make much more complete cocoons [if that is not done]. Besides, they much prefer the leaves of white (4) mulberries to red, since they are milder and sweeter. The size that a silkworm will attain by the end of its salad-feeding period is that which can be seen just above the eggs at the bottom of the copperplate. Now if it has been conscientiously fed with the new food of mulberry leaves, it will gradually grow and eventually reach the large size of the one resting on the green leaf. From its head, marked with spots that look like eyes, though it cannot see, it has eight segments, and on each segment little black rings on either side and finally two little legs. It bears a straight line down its back, greenish in appearance, and has a gray mouth. Its legs, including the claws, are sixteen in number—namely, six small claws under its head, and after two segments

---

3 Very small larvae were fed tender lettuce, which would have been easier for breeders to obtain year round than mulberry leaves.

eight, then after two more [segments], i.e., at its hind end, two more round legs. The droppings they produce are green at first, then dark, as seen lying behind the large caterpillar on the green leaf. The pushed-off skin lies between the mulberry leaf and the lower, complete cocoon; above it lies an opened one, in which the tip of a date seed can be seen, with a husk lying next to it. But the actual date seed is the one on the [opposite] end of that row. The insect above this date seed (with the many eggs underneath it) is the female; the male is the one next to her. This, then, is the clear and true description of the highly esteemed silkworm.

Chapter 1: *Bombyx mori* (silkworm) and *Morus alba* (white mulberry)

Merian chose the silkworm as the first entry after her preface, deeming it the "most useful" and noble of the caterpillars, as well as the best known. Beginning in China, the silkworm moth has been cultivated by humans for several millennia, and the caterpillars and their valuable product were mentioned by Aristotle and Pliny. Silkworms have been a domesticated species for so long that they can no longer exist on their own in nature, and the life cycle of this insect may have been one of the earliest to be recorded in Europe. Illuminated medieval manuscripts show cocoons being gathered and the silk being processed, and as previously described, Renaissance naturalists wrote about the life cycle of this species.[4]

Expanding on the information contained in these earlier images and, like Georg Flegel (Figure 30 in Chapter 2), incorporating part of the mulberry plant, Merian portrayed many essential details of the silkworm's developmental stages in her composition, and further described this life cycle in her text. Some information is present only in the image in Plate 1, such as the arrangement of the hard-sided pupa inside the silken cocoon and the smaller bodied male extruding its meconium.[5] What Merian omits from this image and all others is interesting in itself; she never showed mating insects, something that she certainly would have observed, and that was depicted in Aldrovandi's work as well as Flegel's. However, she is not shy about stating that the adults mated almost immediately after emerging from pupation. As in her preface, Merian is clear that the eggs laid over a period of several days are a result of mating between the adult insects. The physical differences between male and female silkworm moths as depicted and described would have been well known in her time, but as we shall see in later chapters, she was notable in showing sex differences in other species of insects in her books. The adult male and female of most animals exhibit external differences, and in some lepidopterans, sexual dimorphism can be very striking, causing them to look like they could be different species. In other cases, sex differences are less obvious to the casual observer, appearing as in the silkworms as differently shaped abdomens and antennae.

In Plate 1, the image is anchored by the large caterpillar in its last stage before pupation. Each distinct stage of the caterpillar between molts is called an instar, and in lepidopterans, the number of instars usually range from five to

---

4 Pp. 24–30 in this volume.
5 The reddish liquid meconium consists of waste products from larval tissues. Merian does not address this substance in her text about the silkworm.

seven. The final silkworm instar is shown resting on the leaf of a white mulberry plant, their preferred food, and the frass (feces) of the animal is shown on the same leaf. Merian shows more stages of the silkworm larvae than for any other insect, beginning with tiny caterpillars hatching from eggs in the lower corner. Like all insects, the protective 'skeleton' is on the outside, and this exoskeleton must be molted between every growth stage of the caterpillar. Between the cocoon and the largest caterpillar, Merian illustrates the shed exoskeleton of a silkworm, which she terms a "pushed-off skin." Her text adds details regarding the segments of the larvae, and the number and position of their 'feet' and claws. Several views of the pupa and cocoon make their structures clear to the reader, and her text expands the information conveyed with a brief account of the emergence of the adults. The five instars portrayed here, as with all of the insects in her plates, were engraved very close to actual size, an additional source of information in her images. In more detail than for any other species in her books, Merian includes the specifics of silkworm husbandry. She recounts the need to feed the early instars 'salad' (lettuce) when mulberry leaves are not yet available, and methods for keeping the larval enclosures clean. The effect of temperature on their development is noted, an important environmental consequence that is a recurring topic in later chapters.

The silkworm chapter is unusual in Merian's books on moths and butterflies in that she seems to be writing of others who raised the insects; if she raised them herself, this is not clear from her writing. Various biographies of Merian have included the idea that in 1660 at age 13 she began her lifelong study of metamorphosis by raising silkworm moths, but upon further examination, I believe that this interpretation is incorrect. Because silkworms did not exist in nature by her time, it would have been necessary for someone who cultivated them to have given young Maria Sibylla the insects to raise, and instructions on the meticulous process. While this is not impossible, she never directly states that she cultivated the silkworm, and the entries in both her study journal and in her book refer to the process entirely in the third person, as opposed to all other entries, which are written in the first person. In neither her handwritten nor published text is there any indication that she handled the insects extensively, only that she is reporting on well-established techniques for their husbandry. It seems likely that she learned about the process from people raising silkworms in Frankfurt, possibly Jacob Marrel's brother, who may have been in the business of silk production.[6] The potential misconception that this was her first experience in raising insects comes in part from the closing section of her

---

6   Davis, *Women on the Margins*: 143.

first study journal entry. This statement is set off in a separate paragraph after the description of the silkworm life cycle and reads:

> I began this investigation, praise be to God, in Frankfurt in 1660 and offered the first part, from Nuremberg, at the fall [book] fair in Frankfurt in the year 1679; the second part, while residing in Frankfurt, in the year 1683.[7]

It seems more likely that she is referring to the larger work in this notation just below her first study journal entry, and not specifically to the study of silkworm metamorphosis as has been inferred. The investigation that she speaks of resulted in two books, published in 1679 and 1683, so she seems to be speaking of her broader study of metamorphosis. Perhaps a more significant clue is a short note written in her hand on the top of next page of her research notes. Just above her second entry in the study journal an inscription reads "Anno 1660 in Frankfurt am Main, my first observation."[8] This indicates that the small tortoiseshell butterfly pictured below the notation was the first species that she raised on her own (see Chapter 44 and commentary on p. 329).

Thus, an alternative possibility to the conventional assumption that Merian raised silkworms at the age of 13 is that at some point she studied silkworm husbandry as practiced by others, and used these insects as her models. It is possible that as she became more experienced she did go on to raise silkworms. Indeed, it seems unlikely that the detailed image in her study journal, which includes the frass of the caterpillar and the shed exoskeleton (Figure 61), were made by a 13-year-old studying her first metamorphosis. Further complicating the mystery of Merian's early studies is the fact that the finished plate of the silkworm contains much more detail than the small study painting in her journal. Comparisons of the images and handwritten text in Merian's study journal to the engraved images and printed text in her 1679 publication will be made in the commentary on several of Merian's chapters that follow. Beginning with the silkworm journal entry,[9] we can see that there is not complete congruence between the entries, but most of the salient details are the same in both sets of text. In many cases, the phrasing of sections of her writing is identical in both places, but in the silkworm chapter as in others, the book has more details on appearance, in particular color. As mentioned earlier, these physical

---

7   Appendix, study journal entry 1.
8   Appendix, study journal entry 2. *Aº 1660 In francfort am meine erste unterfindung.*
9   Appendix, study journal entry 1.

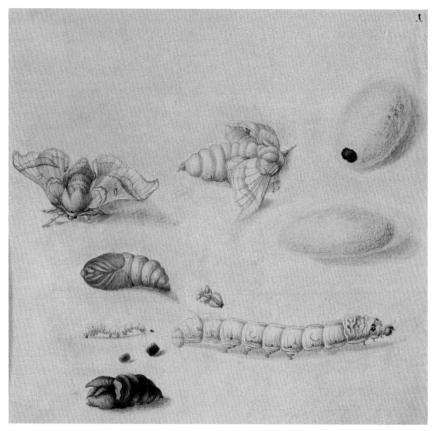

FIGURE 61   Silkworm life cycle in the first entry of Merian's study journal. She painted all
             studies with watercolor and gouache on vellum. Several details shown in the final
             plate are absent from this study.

descriptors are printed in a way that makes them stand out for the reader by using a larger font. The published text on the silkworm contains more information on the husbandry of the insects than does the study journal, and she states her reason for including the silkworm in the frontispiece or title page of her book.

PLATE 2   Merian, 1679. *Raupen*

11. Purple tulip   *Tulipa purpurea*

(5) The present copperplate shows us a lovely tulip that is highly regarded by flower lovers and known as the Marbled Jasper, faithfully painted from life with its white basic color and beautiful streaks of crimson. Before the caterpillar resting on it had grown to this size and full development, I assumed that it fed on the leaves of the plant. However, it would not eat the green leaves, but only the blooms; although besides them it would readily consume the flowers of *Auricula ursi* as well. From April onward I maintained the young caterpillar with these plants, which it destroyed completely and, to all appearances, found much to its liking. After it had reached the size and form of which I speak: dark below, but a light wood color with cross-markings above, it lay down, very still and as if resting, and turned into the shape of a date seed, appearing to be completely dead. But as soon as it was placed on a warm hand, it began to move, and one could clearly see that in such a changed caterpillar—or rather, in its date seed—(6) there must actually be life. Now it is no small wonder that from such date seeds, were one to cut them open not long after this change, nothing but a colored, watery material would run out; while by contrast, at the proper time, such a changed thing comes forth, namely a new insect, often gloriously adorned and having the most beautifully colored form. On the arching green tulip leaf the date seed, which was liver-colored, is shown. But in May a moth-bird came forth, like the one in descending flight, which likewise had a liver-colored head. Its two forewings, those closer to the head, were of the same color, but its body and two hind wings were gray or silver-colored. The rest of its attractive markings are all of blackish hue, as the aforementioned moth in the center clearly shows.

Chapter 2: *Diarsia brunnea* (purple clay moth) and *Tulipa gesneriana* (tulip)

The second chapter of the *Raupen* book is about one-third the length of the first, and the clay moth entry is more representative of her usual level of detail than the silkworm description. Merian makes it clear that she raised the clay moth larvae on tulip flowers, writing that the caterpillars disdained the leaves for the petals. In this chapter, she addresses food choice in a native European moth species for the first time, although she shows it on a garden plant introduced from central Asia. The tulip featured here was a highly valued variety of an early spring bulb, the Marbled Jasper, which produces a white and red/purple variegated blossom. However, Merian explains that these caterpillars were also very happy to eat the flowers of a type of primrose (a native plant) to the point of their complete destruction. In fact, these caterpillars feed on a wide array of plants, what ecologists would term a generalist feeder, as opposed to species that are specialists, feeding on only one type of plant. The choice of a tulip as the first flower in the book may have been a tribute to her stepfather and teacher, Jacob Marrel, who was a respected painter of tulips. In addition, the tulip would have been one of the earliest flowers to appear in Nuremberg gardens, fitting with the overall sequence of plants and insects appearing in the book. Study journal painting 69 (Figure 62) includes a smaller instar caterpillar that Merian omitted from the finished image, another example of her compositional editing process.

Merian recounts the changes undergone from caterpillar to pupa to adult, a pattern of descriptions that she established in Chapter 1 and would continue to use in all of her writings on insects over several decades. Her information on the timing of these changes usually corresponds very well with what is published in modern scientific literature for each species. Here she notes that the clay moth caterpillars emerged in April and the adults in May. Various general observations are inserted throughout her chapters, and in Chapter 2, perhaps the most notable is her account of the pupa, which although seemingly lifeless, would move when held in a warm hand. This evidence of life was a "wonder" to Merian, because when she investigated further by dissecting the early-stage pupa, she found only watery material inside. She is careful to state that this was a recently formed pupa, and later in her book (Chapter 5), she describes the components of a pupa in a later stage of development. Once again and as we see in all chapters, Merian includes color information such as light wood-colored, liver-colored, silver-colored and blackish hue.

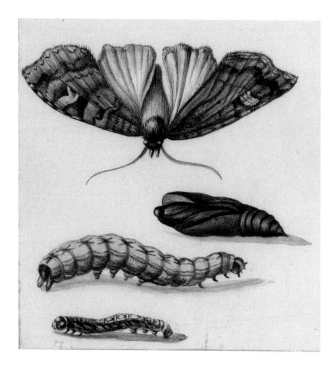

FIGURE 62
Merian. Study journal
painting 69. *Diarsia
brunnea* (purple clay
moth)

PLATE 3   Merian, 1679. *Raupen*

III. Blue elder   *Sambucus cærulea*

(7) Next in our sequence is this pretty, light-green caterpillar dotted with tiny specks; it has a yellowish-green stripe on its lower body and two small black horns on the front of its head, and it moves very nimbly. I have found ones like this in April on blue lilac—or blue elder, as it is called hereabouts—which they take as food; and here I have placed one on a green leaf of that sweet-smelling flower. Now before it changed, or was about to turn into something different, it pushed its skin completely off and, at the beginning of May, turned into a date seed tending toward a dark liver color, before which it made a thin white cocoon in which it became the aforesaid date seed. One such, broken open and empty, can be seen on the bottommost green leaf. It lay completely still, unless one placed it on a warm hand; then it would move. At the end of the month of May, it produced a light yellow ochre moth-bird for me, with four yellow legs and two horns of the same color, two brown eyes, and, on its two front (8) wings a definite white stripe and two large brown spots, but whose center was somewhat lighter. This small but charming and attractive moth-bird (which has a very rapid flight) I have showed higher up, resting on one of the little blossoms. The branch of blossoms seen here I painted from life. It serves as evidence that there is nothing to be found in this, my modest little book, that was not first painted and then etched or engraved in copper by me, in order that the work be as complete, useful, and clear as possible when it sees the light of day.

Chapter 3: *Cosmia trapezina* (dunbar moth) and *Syringa vulgaris* (blue lilac)

The manner in which caterpillars move and adult moths fly was recorded by Merian beginning with this entry about the dunbar moth. Another species that is active in early spring, she describes this caterpillar as moving nimbly and having adults that fly rapidly. Even the movement of the pupating insect is described, as she once again remarks upon the motions of this seemingly dormant stage when held in her warm hand. The pupal casing is shown as it would appear after the adult emerged, and attached to the attenuated end is the shed exoskeleton, something that she depicts in several plates and comments on in Chapters 6 and 11. Her delight in the organisms shines through even when describing this seemingly plain moth, which to her is "charming and attractive." An oddity in her description occurs here and elsewhere in the volume, where she denotes the adult insect as having four legs, when in fact they would have six legs, as is the case for all adult insects. In her study journal image on vellum three legs are showing (Figure 63) and in the engraved copy of the dunbar moth, four legs are in evidence; the hind legs are obscured by the wings. In some insects, the two front legs are very much reduced and are tucked up against the body, so it does appear as if they have only four legs.[1]

Merian's title refers to the spring-blooming plant by the common and Latin name for the elder, but her text also mentions the lilac. The plant is indeed the blue lilac, *Syringa vulgaris*, another plant introduced into northern Europe, which has a very different flower structure than the elder. The taxonomy of plants was much further developed by the 17th century than that of insects, which was almost non-existent. Local 'common' names for plants were used by Merian and while these are of historical interest, they add confusion to a naming system that was far from standardized in the pre-Linnaean era. As discussed earlier, she may have used a variety of sources to supply the Latin plant names. However, if she used the common name 'elder' as the basis for this plant, this would have led her to call this plant Sambucus in error.

Merian's final point in this entry was critical in establishing her authority over the subject matter and her credentials in producing such a book. Many natural history books contained the disclaimer that organisms were 'from life,'

---

1   Four-legged adults also are mentioned by Merian in Chapters 11, 19, 26, 39, and 44. In about half of her chapters, adults are described as having six legs, and in the remaining chapters, leg count is not included in the text.

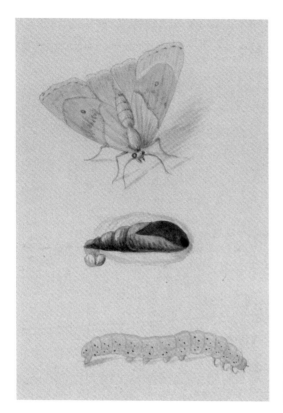

FIGURE 63
Merian. Detail from study journal
painting 80. *Cosmia trapezina*
(dunbar moth)

but as discussed earlier, this was not always the case.[2] Here she asserts that
every image in the volume is her own work, from painting to the finished plate,
adding that this was intended in order to achieve clarity and completeness,
as well as to create a book that would be useful. The statement adds further
evidence to counter scholars who believe that her husband may have assisted
with the plates. Merian makes a statement quite to the contrary here, noting
that she has painted the moth and the blossoms, and furthermore that she has
etched and engraved all copper plates in the book.[3]

2 See pp. 24–6 in this volume.
3 Most of the lines in her plates appear to have been etched, but she may have used engraving
   techniques in some areas as well.

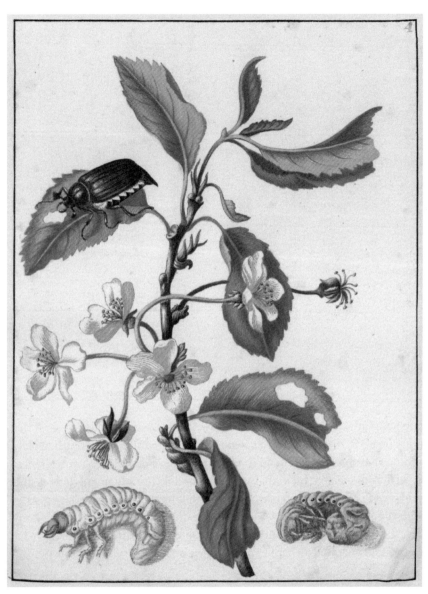

PLATE 4   Merian, 1679. *Raupen*

IV. Black sweet cherry, in flower

Cerasus nigra, dulcis, florens

(9) This curious worm is found in the earth and, while unknown to many, it is not unfamiliar to the countryman and farmer, as it is they who till the soil; they know it as the *Engerling*. It is very pale yellow in color and looks like a maggot or cheese worm, except that its head is more yellow. Roots in the ground provide its food. I have found it, very small, in October but, in March and April, fully grown and of a size like that shown here at the bottom of the copperplate. After it has attained its full size, it will crawl off somewhere in the garden where it can remain lying quietly; then, all curled up, it enters its change, as can also be seen at the bottom along with the other one. Toward spring, that is in April or May, it becomes such a large beetle as I have shown at the top sitting on a green leaf, having six legs and a lovely light brown color but with a black-and-white underbody. The female has small horns while the male has broader ones. They return underground to lay their eggs, from which then the same kind of worms will come. In addition, I have painted sweet black cherry blossoms (10) from life and shown the beetle on them, since these are what they prefer to eat, although they also feed on many other kinds of herbs and tree leaves. Thus the larch, as well as the chestnut and the great maple, will be the first ones devoured by them. Whether these beetles crawl back into the earth I do not really know; but it does appear that this is what happens, especially since I have discovered them in the ground at various times. One will hardly ever see the worm crawling above ground by day, but surely at night, when it feeds on such of the above-mentioned things as it finds.

Chapter 4: *Melolontha melolontha* (cockchafer) on *Prunus avium* (sweet or wild cherry)

The cockchafer is one of the few non-lepidopterans to be featured in the lead role in a plate of Merian's *Raupen* books. Unlike moths or butterflies, these beetles are voracious herbivores even as adults; to emphasize this she shows the adult making a hole in a leaf, and she includes another well-chewed leaf lower on the plant. As she states, the adults feed on leaves from a wide variety of trees, whereas the larvae consume plant roots. The sexual dimorphism in the adults is described; as in many species of insect, the male has more well-developed antennae ("horns" in her words), the better to locate the females using chemical cues. She also notes that the cockchafer eggs are laid in the ground, and that she found both adults and larvae in the soil, further indication of the fieldwork that she undertook in her studies. She found beauty even in these very common beetles, calling them a "lovely light brown."

The comparison of the beetle's larvae to the well-known "cheese worm" is a visual reference for her readers. Merian writes about the larva curling up and entering its change, but she does not refer to a date seed. She understood that the beetle's metamorphosis was different from the transformations of caterpillars, and she depicts the pupating beetle with accuracy. Unlike in lepidopterans, the appendages of beetles can be seen in the pupae, and she provides this information in her usual straightforward manner and without elaboration.

PLATE 5   Merian, 1679. *Raupen*

v. Blue / Oriental hyacinth

Hyacinthus Orientalis, cœruleus

(11) This caterpillar, black on its hind end, having long, black hairs on its back but yellow ones on its underside and head, looks as if it had been decorated with pearls. Many hairs extend from each little pearl, so that the caterpillar is entirely covered with hair. Below the front of its head, on each side, it has three small legs; in the middle of its body, on each side, four legs; and at the hind end, one leg on each side. They eat greedily, feeding on all kinds of flowers and herbs, such as grass, deadnettle, stinging nettle, and the like; also hyacinth, a stem of which is pictured opposite. They also move very rapidly. Gardeners call them *Beermutter*, as if perhaps to say *Perlmutter*,[1] since they appear to be adorned with so many pearls. If one touches them, they roll themselves up in a ball and remain like that until they no longer feel anything; then they gradually stretch out again and crawl away. I have had small caterpillars of this type that attached themselves to a wall or wooden surface in April and became quite hard. One such small caterpillar (12) can be seen crawling on a blossom. The hard, withered stage is very small, however, and is pictured lying next to the large date seed. They remained this way for ten days; then from some of them a completely black fly came forth, but from others, a fly with some yellow in the middle of its body. Now before the caterpillars attain their full size they push off their skin several times; but when they are fully grown they enter their change, losing all their hair beforehand, and make a gray cocoon in which they become a date seed, entirely black. One such can be seen above the caterpillar, without its cocoon and broken open, since the adult has come out of it. But if the topmost skin is peeled away before they come forth, one can see how the insects are lying inside, as the image at the bottom, opposite the caterpillar, shows. For fourteen days they remain like this, undergoing their change, after which the mottled adult moth at the top emerges, a lovely bright red marked with black spots and stripes on its body and hind wings, but with a brown head and forewings, which are also marked with white strips. They have six red legs and two brown-and-white horns; and they produced pale sea-green eggs resembling grains of millet.

---

1   Literally, mother of pearl.

Chapter 5: *Arctia caja* (garden tiger moth) and *Hyacinthus orientalis* (common hyacinth)

The scene of the colorful garden tiger moth on the hyacinth is an odd combination of graceful beauty in the upper portion and an almost clinical presentation of insect forms in the lower part of the plate. The evident contrast reflects a tension in Merian's work between her aesthetic sensibilities and her desire for exactitude in presenting what she observed. A great deal of information is packed into the image and text for this chapter, and much of it closely follows the description for this insect in her study journal.[2] The garden tiger moth's pupa is the second one for which Merian describes her attempts at dissection, but for this study, she waited until just before emergence of the adult and was able to represent the moth as it appears in its final stage of development inside the pupa.

A sinister element is introduced here in the form of parasitic wasps, mortal enemies of many caterpillars. At this early stage in her insect research, Merian, like other 17th century naturalists including Johannes Goedaert and John Ray, was confounded by the disruption in the usual metamorphic cycle when these insects emerged from the body of a caterpillar or its pupal form. The parasitic, or more properly parasitoid, insects lay their eggs in larval moths and butterflies.[3] The developing fly or wasp larvae benefit from this rich source of nutrients, and in the process, destroy the caterpillar or pupa as they develop and hatch. In the 1679 *Raupen* book, Merian refers to these occurrences as false transformations or changes, and she carefully recorded what she observed in these instances.[4] The small Braconid wasps depicted above the pupa in Plate 5, like other insects in her work, are shown roughly life-size and properly scaled to the stages of the tiger moth.

As is typical for most entries, Merian commented on the feeding habits of this ravenous caterpillar, which consumes a wide variety of plants. As usual, she is correct in these observations and in the rest of her description of the garden tiger moth. Chapter 5 introduces the first of several observations on what would now be called defensive behaviors of caterpillars, which serve as

---

2 Appendix, study journal entry 25.
3 Parasites do not typically kill their hosts, but parasitoids do. Common groups of insects that lay eggs in lepidopteran larvae include Braconid wasps, depicted here and in Plate 45, and Tachinid flies, seen in Plates 26, 33, and 49.
4 Other examples of parasitoids appear in Merian's Chapters 5, 7, 16, 22, 26, 45, and 49.

a rich source of food for predators as well as parasites. She comments on the long hairs of the caterpillars, but does not mention here or elsewhere that such defenses sting or irritate potential predators and parasites. Urticating hairs, named for their nettle-like properties, are a common defense in various arthropods, and there is little doubt that she would have known this as often as she handled them. Merian described the behavior of the larvae as they roll into a ball when touched, and this and other defenses are recurring themes in her work. However, she never overtly connects such behaviors to predation or other threats to the caterpillar, which seems odd in light of the fact that she had access to Johannes Goedaert's work. He mentions the stinging hairs and rolling behavior in various caterpillars and went on to discuss predation by birds on larvae.

The life stages of insects portrayed in the preceding three plates are almost identical to images in Merian's study journal. Unlike Plates 2, 3, and 4, Plate 5 shows a number of differences from the paired study image (Figure 64). Minor differences appear in the arrangement of the eggs, and the adult moth shows four legs in the plate compared to two in the study; however, the legs show much more structural detail in the finished plate. Plate 5 has other significant changes from the study: an additional smaller instar of the caterpillar, and a second parasitoid in the form of another wasp. Both of these wasps (referred to by Merian as "flies") are mentioned in the text of her study journal entry for the tiger moth, but only the solid black one appears in her watercolor study image.[5] She added the younger form of the caterpillar to the plate to match her text recounting the fact that the smaller caterpillars are the ones that become hard and withered, producing the "flies." On the other hand, the curled caterpillar in the appended study image is absent from the plate, although she mentions this behavior in her text. It has often been stated that Merian arranged her engraved compositions of insects and plants by copying from her study notes, but as previously mentioned, the reality is more complicated. As with the example cited in Chapter 1 on the silkworm, Merian decided that the original study did not convey all she wished to show in order to complement the text. She often learned more about an insect between the point at which she made the original study and when she designed her finished composition for the plate.

---

5   Merian rewrote her early study journal entries after publishing the *Raupen* book in 1679 and may have added to her original entries.

FIGURE 64   Merian. Study journal painting 25 of *Arctia caja* (garden tiger moth) with
parasitoid. An added piece of vellum shows the garden tiger caterpillar in its
stereotypical defensive position (not depicted in Plate 5).

PLATE 6   Merian, 1679. *Raupen*

VI. Sweet spearwort (buttercup) *Ranunculus dulcis*

(13) Ones of this kind I have always found in the grass where there are many yellow spearworts or, as they are generally called hereabouts, buttercups; which they like to eat, and on which I have kept them for several years from April to the end of May. In the event they did not have these plants, they were also likely to eat sorrel, deadnettle, marsh marigolds, and gooseberries; but as soon as they got their preferred plants again, they would ignore the others. This caterpillar is black with golden-yellow hairs over its entire back, but ones the color of egg yolk below; it also has small white stripes and yellow ones under these. Behind and below its black head there are three pairs of small claws or legs, four pairs in the middle, and two pairs toward the back, as shown at the bottom of the copperplate. If they are touched, they roll themselves up in a ball and remain lying that way for a time until they resume feeding. At the end of May, they looked for a place to retire and made a thin white cocoon, in which they then pushed off their (14) skin completely and turned into a brown date seed such as hangs, together with its cocoon, from a blossom and a leaf opposite the caterpillar. This date seed stirs slightly as long as there is life in it; but mostly it lies quite still, and that for two weeks. Then a very pretty moth comes out, as shown in descending flight at the top. Its head and body, as well as its two forewings, are sulfur yellow with black flecks; its eyes are black; the two hind wings, however, are a lovely bright red and bordered by black and yellow flecks. At the front of its head are two brown horns, and it has six red legs. These moths rest quietly together all day until evening; then they fly about until dawn approaches. Afterwards, they move off to dark places and remain there quite still again. Now since I had many of them together for a number of days, they laid their eggs all together, too many to count and yellow in color, as some can be seen next to the caterpillar at the bottom.

Chapter 6: *Rhyparia purpurata* (purple tiger) and *Ranunculus* sp. (buttercup)

The purple tiger moth has a similar defense system to that of the garden tiger moth, and again Merian describes the larval hairs and the caterpillars' behavior when disturbed. As she shows us, the final exoskeleton when molted remains attached to the pupa, and she establishes that pupation occurs at the end of May. She adds that the adults emerge in June and hide in dark places during the day. As in the preface, she mentions the night activity of moths, including the purple tiger as an example of this. She reiterates this point in several more entries, clearly delineating the nocturnal activity of moths from the day-flying behavior of 'summerbirds' (butterflies).

In this entry, Merian begins to discuss more fully the food choices of larvae, which is a hallmark of her work and of plant-animal ecology to this day. Although the purple tiger caterpillars prefer the buttercups depicted in this plate, she finds that they will eat several other species of plants. Caterpillars can easily consume more than their own body weight in food in a day, so it is not surprising that they may exhaust their preferred food supply very quickly. Her statement that they will ignore other plants and return to the preferred food when it is made available is an insightful ecological observation, and it is a theme to which she returns later in her book.

The image of the purple tiger moth's metamorphosis in Merian's study journal serves as a close model for the engraved plate. However, the published text contains much more information than in her handwritten notes, describing the movement of the pupa and aspects of the adult moth's behavior. Such differences between her study notes and the finished work occur elsewhere, and leave us with a bit of a puzzle. Perhaps she had another set of notes for which we have no record, or she simply may have remembered various characteristics of species that she knew well. At the end of this chapter, Merian remarks on the large number of eggs produced by the female, another observation not included in the study journal. Only a small fraction of the tiny caterpillars that hatch from eggs make it to metamorphosis, most falling prey to predation, parasites, disease, starvation or other hazards. The high number of eggs produced by adult insects is necessary to offset this attrition in order that some survive to reproduce the next generation.

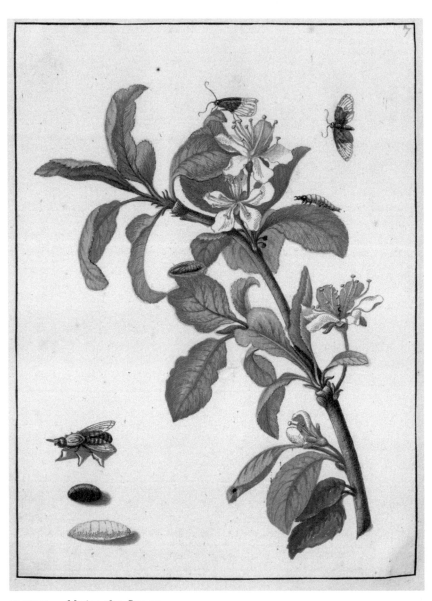

PLATE 7   Merian, 1679. *Raupen*

VII. Flowering plum   *Prunus florens*

(15) Where this kind of moth-bird lays its eggs I have not been able to discover, even though I have located those of many other insects. But I have found its little caterpillar, a most unusual kind, very small, dark green, with black spots and a black head, on the blossoming plum trees in early May. One can be seen here at the top and to one side, resting on a leaf of that tree. When they are touched, they retreat at once, or lower themselves to the ground by a thread from their mouth, then climb deftly back up the same thread to their house, which consists of small rolled-up leaves, one of which is shown below [at the bottom of the stem]. Thus they are concealed from view and are protected and safe from the heat of the sun and the wet of the rain. When they wish to eat, they run out, and as soon as they have eaten their fill they hurry back in; there is a constant, lively running back and forth. Once they have reached their full size they change—inside the same leaf—to a date seed which is half yellow and half brown; there is also a thin cocoon around it, which is pictured along with the date seed opposite the caterpillar and a bit (16) lower, resting on a leaf. Now at the end of the same month, moths or *Schaben*[1] came forth, like those pictured at the top of the plate, one resting and one in flight. Their color is light brown except for the back half of the two outer wings, which is lighter still. They also have six brown legs and horns.

In the lower part of the page lies a maggot-worm, bright yellow, which I have found in all sorts of caterpillar droppings, in which they grow and on which they feed. After three days, it changed into a brown capsule like the one pictured next above it. But after fourteen days, there came forth from the capsule the third, or uppermost, one of these three: a large fly with six black legs; the body and other parts were brown, with a reddish head, as its appearance can be seen in the copperplate.

---

1 *Schabe*: southern German dialect name for moth, not to be confused with the generally understood meaning, cockroach.

Chapter 7: *Hedya nubiferana* (marbled orchard tortrix), Phorid fly (scuttle fly) and *Prunus domestica* (plum)

The larvae of this tortrix moth are the first of several leaf rollers described by Merian, and they are the first caterpillars that she describes as lowering themselves by a thread. She shows the caterpillar on a plum tree, which the species eats along with several other trees, as well as flowers such as roses. The defensive behaviors noted here fascinated Merian, who used vivid language to describe their lively movements up and down spun silk threads and in and out of their leaf shelters. She also documented this use of a 'lifeline' by two more Tortricid caterpillars in Chapters 19 and 24. Goedaert noticed this behavior as well and wrote about it in entry 28 in his second volume, although it is very difficult to identify the type of insect depicted in his image.[2] Both of these early naturalists were describing an important defensive behavior that has not been well studied even in this century.[3]

Leaf-rolling behavior has been investigated to a much greater degree, and is of interest as a defensive and sheltering mechanism for the caterpillar, but also from a community ecology perspective. The shelters created by the leaf rollers also provide microhabitats for other species, thereby bringing even more diversity of organisms into the ecological community living on one plant.[4] Merian writes about other leaf rolling caterpillars in Chapters 24, 28, and 48. From the information here and in many of her entries it is clear that Merian spent countless hours observing the behavior of insects in nature.

The last part of this entry is about a fly whose maggot (larvae) she has seen in "all sorts of caterpillar droppings." She remarks that they grow and feed on these droppings. While it is not possible to identify with certainty the fly from Merian's illustration or her written description, many dipteran larvae do feed on insect frass or on fungi that grow on the frass.[5] She mentions a "common fly" emerging from a pupa whose maggots grew on the dropping of peacock butterfly caterpillars (Chapter 26), and she details the same phenomena in

---

2 Goedaert, *Metamorphosis Naturalis*, Volume 2, Chapter 28.

3 See a survey of the use of silken lifelines by caterpillars in Sugiura, Shinji and Kazuo Yamazaki. 2006. The role of silk threads as lifelines for caterpillars: pattern and significance of lifeline-climbing behaviour. *Ecological Entomology*, *31*(1): 52–7.

4 Martinsen, G. D., K. D. Floate, A. M. Waltz, G. M. Wimp, and T. G. Whitham. 2000. Positive interactions between leaf rollers and other arthropods enhance biodiversity hybrid cotton-woods. *Oecologia*, *123*(1): 82–9.

5 Scuttle flies such as *Megaselia giraudii* and *M. rufipes* are examples of the family Phoridae that breed in the droppings of caterpillars. Disney, R. H. L. 1994. Larvae. In *Scuttle Flies: The Phoridae*. New York: Springer: 34.

FIGURE 65
Merian. Detail from study journal painting 41.
Phorid fly (scuttle fly)

Chapter 33 and 44, again designating the adults as common flies. She thought this pattern of fly generation of sufficient importance to include it in her preface as well. Merian's language is, as usual, straightforward; she describes what she sees and leaves it to others to interpret. It is significant however, that she does not invoke spontaneous generation to explain the appearance of the flies in the droppings. She also separates the generation of this type of fly from those that emerge from the bodies of caterpillars, which she discusses elsewhere in her book. The fly is described and pictured in study journal entry 41, whereas the moth appears in entry 66, so it is not apparent that there was any association between these two species of insects. Merian painted the life cycle of the fly with the same care that she did for that of the moth (Figure 65).

PLATE 8   Merian, 1679. *Raupen*

VIII. Dandelion   *Taraxacon*

(17) Just as this caterpillar has hardly been seen, or little noticed, by many people before this, but now, after observation and inspection, is given closer attention, I am sure that many a person who sees the accompanying little flower, even though he were a great devotee of painting, would not have thought such a modest yellow dandelion with its lacy foliage a fit subject for another's drawing. The caterpillar first appears in April, is entirely brown, and has on its head something like two horns of black hair and on its back five upright, rectangular tufts of hair, which are also black, save that the middle part of the tuft is white on both sides followed by another tuft of black and white, while the rest is nothing but yellow hairs. The caterpillar resting on the bent-over stem of the little flower in the center of the copperplate clearly shows this. Its habit when touched is to roll itself up in a ball and to lie completely still. I have also found it on hedges where there are hawthorns. This caterpillar begins its change in early May, takes its (18) own hair, and also the wood on which it rests, chewing it fine and forming a brown, oblong cocoon out of it, as can be seen lying on the ground at the bottom of the plate. Inside, it becomes a black date seed, which is also covered all over its back with small brown hairs, as can be seen on the ground in the center. It remains resting in the date seed until the end of May; then this curious moth-bird comes out which is gray or ash-colored with yellow flecks. On its body it also has two small tufts of hair and six legs, as the one sitting on a leaf shows. When this insect first comes out, it holds its forelegs, which are hairy, in front of its head all the time so that one cannot see the head, and it remains for several days in that position; only then does it begin to look around. But it flies only at night.

Chapter 8: *Dicallomera fascelina* (dark tussock moth) and
Taraxacum officinale (dandelion)

Another caterpillar with broad food tastes is shown on the dandelion, to
Merian's mind an overlooked plant. Intriguingly, she discusses the "modest"
flower and foliage as a subject for art, and includes the phrase "a great devotee
of painting" to describe a hypothetical reader of her book. Use of the dandelion in this composition may have been a nod to other Nuremberg artists who
used such common meadow plants in their paintings, beginning with Albrecht
Dürer. It is also possible that she was referring to a less well-known Nuremberg
painter, Hans Hoffmann (Figure 34 in Chapter 2). Through her own family
and other contacts such as the Imhoffs, Merian could have had access to such
works and likely was influenced by these master artists. Maria Sibylla again
mentions other artists in the Chapter 36 entry on the plantain.

Merian also refers to the tussock moth caterpillar as being "little noticed."
Ironically, she was honored in 1992 with her portrait on the 500-Deutschmark
banknote, and her image of the tussock moth caterpillar and adult was reproduced on the reverse side of the note. The larvae of the tussock species are
additional examples of caterpillars that roll into a ball when disturbed. The
novel information in this entry is the observation that the caterpillars chew
their own hair along with wood to form their cocoon before pupation. Her detailed image of the eggs shows them covered with a sort of fuzz, although she
does not comment on this covering in her text. The hairy eggs are also evident
in the study journal painting, although there are fewer of them (Figure 66). In
nature, the eggs of the dark tussock are indeed white and covered with a dark,
wool-like substance.

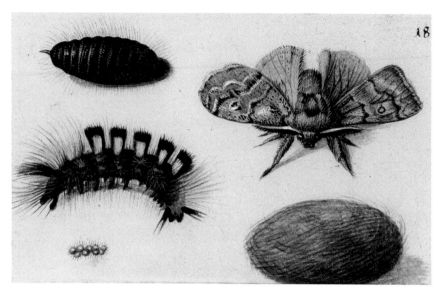

FIGURE 66   Merian. Study journal painting 18. *Dicallomera fascelina* (dark tussock moth)

PLATE 9   Merian, 1679. *Raupen*

IX. A sour cherry with white double blossoms

Cerasa acida, rubra, flore pleno

(19) One often finds very many of this common sort of caterpillar in May, on cherry, plum, pear, apple, and similar flowering trees, all of whose leaves they use as their food. Indeed, they devour so much in a day and grow so fat that they will roll, and soon fall to the ground, as indeed happens. Down their back they have a yellow stripe, and lower down on each side, another of the same color. The rest of the caterpillar is a beautiful green color, the green being distinctly marked with black dots, and on each dot is one small black hair. Beginning from the head, which also has dots, as if two black eyes were to be seen there, they have six black legs, but in the middle of their body eight yellow ones, and at the hind end, likewise adorned with dots, two more yellow legs, as clearly shown on the caterpillar-worm crawling on the green leaf of the double-blossom sour cherry. Then, when they have their full size and are ready to begin their change, (20) they first enclose themselves in a rounded, oblong, white cocoon that shines like silver and is hard as parchment; and inside, bit by bit, turn into a completely brown date seed, one of which can be clearly seen on the curled green leaf. The cocoon, however, is shown resting on the flat leaf with the hole in the middle. After it has remained inside for fourteen days, there comes from its date seed a moth-bird such as commonly flies only at night: The two forewings are reddish-gray, and each wing has two well-defined, light-green spots; the hind wings are of the same color as the forewings, though somewhat lighter. The head and the front half of the body are brownish. Apart from this, everything is shaded brown by Nature's art, just as the insect in downward flight shows.

Chapter 9: *Diloba caeruleocephala* (figure of eight moth) and
Prunus cerasus v. *florepleno* (double cherry, double flower)

Another generalist that feeds on a variety of fruit trees, this rotund caterpillar
is shown feeding on a leaf, and Merian points out that these larvae eat so much
that they get fat and roll out of the trees and onto the ground. Caterpillars are
depicted in various positions throughout the book, but only infrequently are
they shown actively feeding (see others in Plates 35 and 48). She again remarks
on the night-flying activity of the adult and, as in many of her compositions,
features the adult flying down to the plant with wings spread. Whether this
was an aesthetic decision – most lepidopterans are more brightly colored on
the top of their wings – or for identification is not known, but judging from
Merian's other choices in posing insects, the latter seems most likely. It also
could have been a compositional device meant to denote life through motion.

Here, as in Plate 8, she shows the cocoon separately from the pupa, a compositional device employed elsewhere in the book for species that cover the
hard pupal casing with a soft cocoon. Aided by her clear textual description,
the image helps her reader to understand the layered nature of the cocoon
and pupal structures. One also gets a sense of how the cocoon appears in life,
"shining like silver" and "hard as parchment," characteristics that cannot be
depicted through her art. Because her subjects usually were metamorphosing
in captivity, Merian was not always able to observe where their pupation took
place in nature. Some moths pupate attached to vegetation, but many species,
like the pease blossom (Plate 40), the pug moth (Plate 41) and the dark sword-
grass moth (Plate 43), pupate underground. In these instances, she showed the
pupae lying on a leaf or beneath the plant.

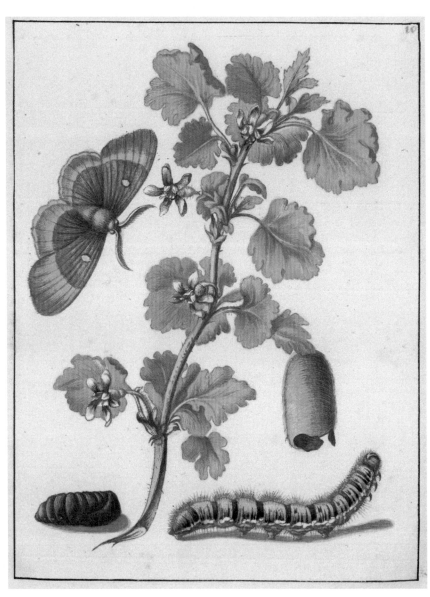

PLATE 10   Merian, 1679. *Raupen*

x. Large / Spanish gooseberry, in flower

Flos Grossulariæ, sativæ, spinosæ

(21) No year can I remember in which there were so many and extraordinarily large caterpillars that did greater damage to the trees than this year of 1679. True, there has been many a year when plenty of these vermin appeared early in the spring season, but then gradually spoiled[1] or quickly came to nothing. This time I have been studying the caterpillars, which I found only singly in other years but now find in large numbers and often in groups. Some of them (let me note well) were twice the size of the half-erect caterpillar crawling at the bottom of the page. What might be the reason for such great numbers, and my opinion on their size, will be explained with Plate XVIII. Now these caterpillars have their start in April, are brown in color, marked with black bands and white spots. They move very slowly and favor the *Stichel-* or *Stachelbeere* [gooseberry], also called *Klosterbeere*, as food. For this reason I have shown a blossoming stem of the large variety, since they fed on it in preference to the others. They eat vigorously and (22) like to have water; for if they are regularly given a few drops, they are refreshed. One must especially do this when they are in danger of drying out, otherwise they will die. They also eat grass and the leaves of plums and roses. In June they make a round cocoon in the shape of a slender, oblong egg, which is quite hard and brown in color; one such is shown open and hanging from a leaf. In it, they become a date seed, the form of which is shown resting at the bottom, next to the caterpillar, its color dark red. In the month of July, a moth like the one flying above came forth from a date seed, the female the color of wood, but the male a pretty dark brown; they also have large horns. As soon as they have crawled out, they mate; after that, the female lays the eggs, which are larger than millet seeds and liver-colored. Furthermore, on each of the two front wings is a round white spot, and its legs number six. It flew only at night, while by day it was quite still, without the slightest movement. This was not surprising, since it is generally true of other moth-birds, too, with few exceptions.

---

1   In other places as well she refers to dead larvae as spoiled.

Chapter 10: *Lasiocampa quercus* male (oak eggar moth) and
Ribes uva-crispa (gooseberry)

Here Merian pictures a sprig of gooseberry in flower, and later in Chapters 25
and 31, she offers two more views of different varieties of gooseberries that
are bearing fruit. This entry on the oak eggar moth describes the exceptional
population boom of caterpillars that occurred in 1679. The date tells us that
Merian was probably finishing the text of this book after the growing season of
that extraordinary year of insect abundance. Not only the number of caterpillars appearing that spring, but their size was unusual, and Merian emphasizes
that fact by noting that the larval specimen shown in plate 10 was only half
the size of others she saw. She writes here that in Chapter 18 she will further
elucidate this phenomenon.

The study journal notes for this entry are much abbreviated relative to the
finished text, and the notes refer to the pupae as yellow as opposed to brown
as in the *Raupen* book.[2] The cocoon is painted yellow in both images, but this
difference in color description could be due to the fact that oak eggar cocoons
can vary in color from brown to yellow.

The watercolor painting of the insect's metamorphosis in the journal shows
eggs, which were not included in Plate 10. It is probable that Merian chose to
omit eggs from Plate 10 because she understood that the adult pictured is the
male oak eggar moth. She describes the male moth as darker brown and with
larger "horns" (antennae) than the female; however, her study journal does not
make any mention of this sexual dimorphism. Interestingly, Plate 21 pictures
the female of the same species and the composition does include eggs. It is
unclear whether Merian knew that both of these moths were of the same type
or, as we would say today, species. Perhaps when she was unsure, she simply
omitted mention of any connection; overall, she appears to avoid speculation
in her writings. Nothing in her text with Plate 21 indicates that this is the same
type of moth as the one in Plate 10, even though the larvae are identical, and
Merian described both caterpillars as having a penchant for drinking water.
Another Lasiocampid caterpillar, *Euthrix potatoria*, has the common name of
'the drinker' due to its tendency to seek drops of dew. Her observations of caterpillar behaviors appear in many places throughout the book, and often refer
to their movement, nimble in some cases, or slow, as for the oak eggar larvae.

---

2 Appendix, study journal entry 11.

PLATE 11   Merian, 1679. *Raupen*

XI. A sour cherry in flower   *Cerasus austera, florens*

(23) These small caterpillars, one of which is pictured here crawling down the flowering cherry stem, are dark brown; the first row of rather large, light spots, starting from its back, is white, but the two lower rows of small dots are yellow. These I have found in the month of May on flowering sour cherry and plum trees, on whose leaves they fed. These caterpillars are very quiet by day, not moving at all except when they want to eat. At the end of that month, when they were about to enter their change, they placed themselves on a wall and drew over themselves a white cocoon that had a different shape than usual and looked much more peculiar, like the cocoon shown resting on one of the blossoms. Not long after that, they turned into date seeds; I have removed one and shown it farther down on the receptacle of the aforementioned blossom to make it more clearly visible. At the back of it something dark can be seen, which is its cast-off skin, still clinging to it, as (24) also occurs with many other kinds of caterpillars before they change or are about to turn into date seeds. Now when the insects inside are ready, they chew their way through the date seed and the cocoon and turn into a small moth-bird. This occurs in June, after they have lain fourteen days undergoing their change. They have four legs and four wings; the two forewings shine light gray in the center, but are somewhat darker at their tips and more brown toward the head. The two hind wings are brown in the center; the rest, along with the body, is of that other, light-gray color. The insect in downward flight can be clearly seen on the one side of the copperplate.

Chapter 11: *Nola cucullatella* (short-cloaked moth) on *Prunus cerasus* (sour cherry)

The short-cloaked moth is another that feeds on spring-flowering fruit trees. In Plate 11, Merian presents a beautifully detailed branch of sour cherry to anchor a composition featuring a small and rather plain moth. Again, she pictures the attached "cast-off skin" or exoskeleton on the end of the pupa, and for the first time she describes this structure in her text, mentioning that the outer casing is shed in this manner by various caterpillars. In both the study journal painting and in the engraved plate, she shows the pupa separately from the unusually shaped cocoon that would normally enclose it, doing so "to make it more clearly visible." She writes that this caterpillar moves very little by day, and discusses where and how they pupate, but she does not elaborate on other behaviors of this insect.

The study journal image has some notable differences from Plate 11. In this instance, her painted study of the caterpillar matches the written description and the caterpillar in the painted counterproof does not. This disparity is unusual for this counterproof edition. However, the caterpillar of this species can range in color from light to dark, so the image in Plate 11 does not misrepresent how it can look in nature. The study image also shows a fly and its pupa, and this insect is likely a parasitoid that attacks the short-cloaked moth (Figure 67). It is unusual that she did not include it in the engraved plate as she did for other species of lepidopteran parasites.

FIGURE 67

Merian. Study journal painting 57. *Nola cucullatella* (short-cloaked moth) and an unidentified parasitoid not included in the published plate.

PLATE 12   Merian, 1679. *Raupen*

XII. Yellow wallflower   *Viola lutea*

(25) So that the exceptionally useful and fragrant yellow wallflower, attractive in appearance, might also be presented here, I shall report that although I fed this very slender and delicate little caterpillar with the plant's leaves for several years in succession, I obtained nothing more from it; instead, it dried up. Last year, however, it not only remained alive, but also changed in the natural manner. The little caterpillar crawling upon a turned-over green leaf of this same wallflower is speckled white and green. Underneath and behind its head it has three small green claws, a round leg on its eighth segment, and on each side of its hind end another of the same kind. When this caterpillar moves, it draws its hind legs up to the front ones so that its body rises up in the air, and it looks almost comical; then it takes a step the entire length of itself. Now after I had fed this little caterpillar into June on the leaves of yellow wallflowers, and when it was ready to go into a date seed, it came to rest on a green leaf and soon made a very thin cocoon (26) around itself, in which it then turned into a brown date seed as shown on the lowest green leaf—an active and lively date seed; for as soon as it was touched, it would move about smartly and for quite some time. But at the beginning of July it came forth and turned into a charming little moth-bird that was very lively and nimble in flight. Its color was light brown with dark brown flecks and spots. It also had two black horns, six legs of the same color, and two black eyes, as can be seen on the insect in downward flight whose image appears alongside the wallflower blossoms.

Chapter 12: *Xanthorhoe fluctuata* (garden carpet moth) and
Erysimum cheiri (wallflower)

The wallflower appears to have been a favorite flower of Merian's, and a more
elaborate version with multiple blooms figured in the third set of her *Blumen*
book prints.[1] In the natural history of her *Raupen* book, she implies that she
chose to include this rather non-descript moth because it fed on the plants'
leaves, making the flower the star of the chapter rather than the insect. In
fact, the garden carpet caterpillar is a generalist and feeds on a wide variety
of plants. The inclusion of this decorative flower rather than cabbage or kale,
which it would also eat, was most likely an aesthetic or marketing decision
by Merian, who realized that gardeners were a large part of her audience.
However, the name 'Viola' is given for the plant, which would indicate a violet
rather than wallflower.

In this chapter, she establishes her patience and persistence in learning
about her small subjects by recounting the number of years required to obtain
a successful metamorphosis of the garden carpet moth. This is the first entry to
describe a 'looper' or what some today may refer to as an 'inchworm.' The modern family name Geometridae relates to this unique form of movement by the
caterpillars, which resemble surveyors measuring the earth (geo) with a meter
stick by raising, turning and dropping the stick to the ground.[2] In conjunction
with the description of their movement, she elucidates the anatomical placement of the larval prolegs specific to each taxonomic group of moths and butterflies (she usually calls the prolegs "feet").[3] In the Geometridae, the prolegs
are present in the front and back of the larvae and absent from the middle, as
Merian describes here and again in Chapters 25, 29, 37 and 41 on caterpillars of
this family. She finds the caterpillars' looping movements amusing and the motions of the pupae when touched to be active and lively. Her statement that their
pupae move "smartly" for some time after being touched is significant to pupal
defense against parasitoids,[4] although Merian did not make this connection.
These behavioral descriptions and others such as the "nimble" flight of the

---

1   The wallflower in the sixth plate of her 1680 *Blumen* book nearly fills the plate, leaving little
    space for the addition of an insect's life cycle. Merian, 1680, *Neues Blumenbuch*, Plate 6.
2   Goedaert described this type of caterpillar as "creeping, raising himself, and falling down,
    time and again; like a land-surveyor, measuring land with the pole." Goedaert, *Metamorphosis
    Naturalis*, Volume 2: 58.
3   Caterpillars, like all insects, do have six small legs, but much of their locomotion is aided by
    the stubby prolegs.
4   Gross, Paul. 1993. Insect behavioral and morphological defenses against parasitoids. *Annual
    Review of Entomology*, 38 (1): 255.

FIGURE 68   Counterproof Plate 12 from Merian. 1712 *Rupsen*. The life
              cycle of an unidentified fly was added to the original
              plate etched for the 1679 edition.

adult, further demonstrate how much time and pleasure she took in observing
living insects. Plate 12 was one in which insects were added to the image for
the Dutch edition (Figure 68).

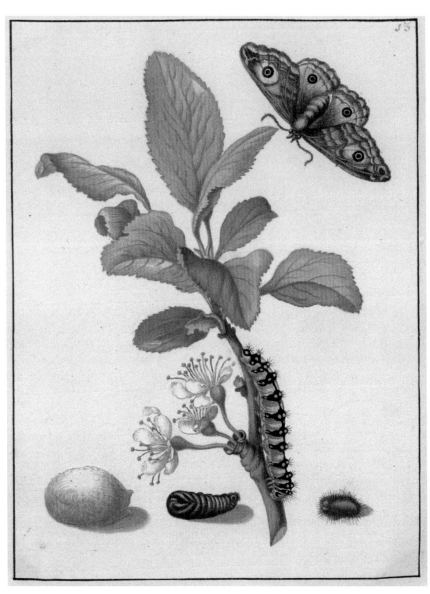

PLATE 13   Merian, 1679. *Raupen*

XIII. Damson plum blossoms   *Flos pruni Damascenæ*

(27) This kind of caterpillar I have had a number of times, feeding them on sweet plum and also pear leaves until the end of June. The caterpillar is a handsome green with a black stripe from its head all the way down its back, and with wider bands running down its sides, though not the entire way down. At the top of these, on each side, are golden-yellow beads, and farther down, golden-yellow spots. At the ends of the black bands, on both sides, a sulfur-yellow line runs the length of the caterpillar; it also has smaller white beads. And from all of these beads, little black hairs stick out. Underneath its head it has three flesh-colored claws on either side, four round legs in the middle and two more on its hind end, as shown on the caterpillar crawling down the branch of the plum. If it is touched, it rolls up completely, like a bale of hay, and remains thus for some time before it uncoils again. Eventually it constructs a hard oval cocoon that gleams like silver, as (28) seen at one side at the bottom. Inside, it becomes a date seed, brown at the front and somewhat reddish at the back, one of which lies at the lower center. It is quite hard and motionless until the beginning of August, when a beautiful moth comes out, as can be seen flying at the top. On each of its wings it has a black-and-yellow ring; the center of the ring is also black; the two forewings are red with brown, spotted stripes; the rest is white. Its head and body, as well as eyes and two horns, are brown in color. Its two hind wings are golden yellow, also brown at the back, though edged in white. It also has six yellowish-gray legs. This moth, like the others, sits quietly by day but flies at night, so that we rarely get to see it. Hence, I spared no effort to find the caterpillar and eventually got the moth-bird by that means.

Chapter 13: *Saturnia pavonia* male (emperor moth) and *Prunus domestica subsp. insititia* (damson plum)

The moth in Plate 13 is the male emperor moth, and in Plate 23 Merian separately depicts the much larger but less brightly colored female. Again, it is not certain whether she recognized these as the same type of insect, but as with the separate images of the male and female oak eggar moths (Plates 10 and 21), no eggs are shown in the etched plate of the male, whereas in Plate 23 Merian does include eggs with the female emperor moth. As with the oak eggar moths, the watercolor studies of both sexes of emperor moth include eggs (Figure 69). Somewhere along the way, she decided not to include eggs in Plate 13, providing evidence that when she made the plate for her book, she understood this adult to be male.

The caterpillar, pupae, and cocoons of both sexes are depicted in much the same way in her text and in both plates, although as seen in below, she shows a different color morph of the caterpillar in her study journal entry for the female. It may be that she puzzled over the relationship between these insects but could not reconcile them as the same type of moth based on the striking differences in the adults, and perhaps she was confused by differences in larval color patterns. Caterpillars of the same species can vary in color depending on the instar described, the food they have eaten, and other factors, making larval identification challenging even for scientists today. In Chapter 23, she mentions the stiffness of the larval hairs, whereas in Chapter 13 she emphasizes the defensive behavior of caterpillars, which curl up "like a bale of hay" and remain still for a while after being disturbed.

Merian recounts her efforts to obtain this beautiful adult moth by "sparing no effort to find the caterpillar." In Chapter 23, she again mentions the exceptional moth and her efforts to raise the caterpillars until her eventual success in obtaining an adult from the pupal stage led to "such great joy" that it was almost beyond her description. In the later entry, she reports that she had the emperor moth larvae for several years and fed them on material from a variety of fruit trees.

FIGURE 69   On the left is a male *Saturnia pavonia* (emperor moth) from Merian's study
journal painting 77. The larger female of the same species is pictured on the
right and is labeled 199b in the study journal. The eggs, pupa, cocoon and
larvae (top right) are on a separate piece of parchment and numbered 199a.

PLATE 14   Merian, 1679. *Raupen*

XIV. Red currant, in flower   *Grossularia hortensis, non spinosa, florens*

(29) Pictured here are two insects of a single type, one of which is seen in flight, the other resting; and which, to distinguish them from those shown up to now, I always call summer-birds. For even though those insects which I have called moth-birds also fly in summer, one does not so easily catch sight of them, since they prefer to fly in the evening and at night rather than in daytime. They also tend not to fly too high in the air. The handsome summer-birds, however, prefer to fly in daytime and also like pleasant, cool, fair weather and fly very high and rapidly; this type can flutter upward as fast as any summer-birds. I found their caterpillars in May on currant bushes, of which the [fruit] clusters and leaves are their food, as one can see in the illustration. They also eat stinging nettles and gooseberries. On the entire hind part of their body they are yellow, (30) but on the front half they are white on top and yellow underneath. The top half of their head is black and white. From beneath, at the end of each segment, they are shaded with black halfway up. From the center of the segments, white, branched hairs sprout, as can be seen on the caterpillar coiled up and hanging from the stem at the bottom. Once they have attached themselves this way to a wall or wood surface, they change within a day into a date seed looking very much like a human face and seeming to be made of gold and silver. Thus it gleams from below when it has attached itself. At the end of June, the summer-birds mentioned above come out. On their inner sides, [the wings] are golden- or saffron-yellow with brown spots; on the outer sides, however, they are all brown and have a clearly visible small, white Latin 'C' on them. Hence I have generally taken to calling them "the C." They are altogether very nicely speckled with larger and smaller spots, with the very same brown color, as one can see clearly on the fresh, just-emerged insects. They have four wings, four brown legs, two long and two shorter horns of the same color, as well as a brown body and head.

Chapter 14: *Polygonia c-album* (comma butterfly) and *Ribes rubrum* (red currant)

This plate is somewhat unusual in her *Raupen* book, showing the adult in two positions, a compositional device she uses primarily for butterflies to show the different color and patterns on the upper and lower wings. As she states, the larvae feed on currants, but she correctly includes the fact that they will also eat nettles. This caterpillar's image was published earlier in her 1675 *Blumen* book, and both that etching and Plate 14 show it in the same posture as in her study journal (Figure 70).[1] Although Merian depicts the pupae hanging from a leaf, she makes clear in the text that they more typically attach to a surface like a wall or other wood. For the pupae of most butterflies, she gives positional information in the images, and in her text, she speaks of them as attaching themselves as they begin this stage of transformation.

The comma is the first butterfly to appear in the 1679 *Raupen* book, and she distinguishes these 'summer-birds' both by their day-flying habits and by their tendency to fly higher than the 'moth-birds' that are mostly active at night.[2] Merian adds some other differences between the two groups of lepidopterans in later chapters. In Chapter 38, she mentions the feeding habits of the adult swallowtail butterfly and in Chapter 43 she reiterates that butterflies have smaller heads than moths. In the same chapter, she adds that at night, rather than flying about, the summer-birds "stay in their accustomed places," a probable reference to the tendency of many species of butterflies to roost in large groups, something that is more unusual in moths.[3] She mentions the swarming behavior of both moths and butterflies in Chapter 43 and elsewhere, describing what were likely mating swarms. The time of day for these swarms varies by type of insect, and she includes these patterns in most of her descriptions, although she does not link the behavior with mating. She also noticed the differences in the minute scales that cover the wings of butterflies and moths, giving them their color, and elaborates on this in Chapter 32.

Merian states that the comma butterfly has "four brown legs," but of course, like all insects, these have six legs. However, as shown in her study journal image and in the engraved image of the resting butterfly, only four are usually on view. Many butterflies have a greatly reduced pair of forelegs, and these are

---

1   The *Blumen* book plate is shown in Figure 43 in Chapter 2.
2   The insect order Lepidoptera includes all families of moths (roughly 160,000 species) and butterflies (about 20,000 species). This relative abundance is reflected in Merian's 1679 *Raupen* book, which contains thirty-six types of moth and only six species of butterfly.
3   Capinera, John L. 2008. *Encyclopedia of Entomology.* Volume 4. Berlin: Springer Science & Business Media: 646.

FIGURE 70   Merian. Study journal painting 26. *Polygonia c-album* (comma butterfly)

often folded up out of sight. Merian noted this trait in the comma butterfly, and indeed, which is in the family Nymphalidae, a group that is known for this trait. She was far from the only person to remark on the crude resemblance of this type of chrysalis (pupa) to a human face.[4]

---

4   Goedaert mentions the appearance of other pupae as looking like a face, but not this species. Goedaert, *Metamorphosis Naturalis*, Volume 1: 4.

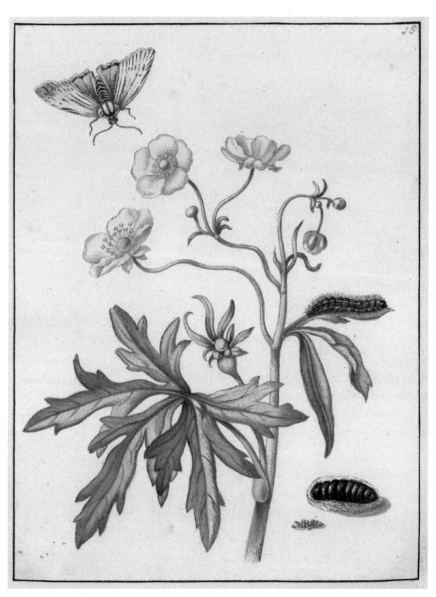

PLATE 15   Merian, 1679. *Raupen*

xv. Wild crowfoot (creeping buttercup)

Ranunculus pratensis

(31) A little black caterpillar like the one sitting on a small green leaf in the middle, to one side of the page, has on its back a golden-yellow stripe and white spots on the black part; also a black head and six black legs. The underbody and the remaining legs, however, are brown. Now I have not only found this caterpillar in April on spotted crowfoot, known in this region as the buttercup, but also maintained them on it. For that reason these little yellow flowers were painted from life by me and placed on view here. However, at the beginning of the month of May it turned into a completely black date seed, one of which is pictured at the bottom. It remained lying in this form into June, when a beautiful moth crept forth, as can be seen in flight at the top. This insect's head is golden yellow, with black eyes, horns, and bands; behind the head it is sulfur yellow, as are its two (32) forewings, and marked with black-spotted bands. The rear part of the body, along with the hind wings, is also golden yellow with black bands and flecks. The six legs are likewise black and golden yellow. This nicely colored insect immediately laid golden yellow eggs that gleamed like true gold and were very small, as can be seen at the bottom, next to the date seed. Whenever this caterpillar is touched, it rolls up and remains motionless, as if dead. It has also fed upon the plant commonly called *Küheblume* [dandelion] or, by those familiar with medicinal herbs, *Taraxacon*.

Chapter 15: *Spiris striata* (feathered footman) and *Ranunculus acris* (buttercup or crowfoot)

In this chapter, Merian noted that she found the buttercup serving as the food for this caterpillar and painted it from life, another statement emphasizing her first-hand empirical observations. She mentions other foods for these larvae, which are indeed generalists on a variety of herbaceous plants. As with several other types of caterpillar these curl up and 'play dead' when disturbed. In her 1679 *Raupen* book, she illustrates eggs for sixteen species of moths that she raised, and in only one case does she describe the eggs of a butterfly (Chapter 45). Many moths lay their eggs soon after emerging from their pupae, making it relatively easy to collect and paint the eggs for these captives, as she did for the feathered footman. In this plate and in Plates 5, 6, 8, 15, she positions the eggs to one side of the plant rather than on the vegetation, and this may be because in captivity natural oviposition sites were not available to the insects.

The caterpillar in the study journal image is smaller than the one in Plate 15, and the cocoon and pupa are combined in the finished plate, whereas they are separated in her study (Figure 46 on p. 87). In an uncolored plate, her etched representation of the relationship of the pupa to the cocoon that encases it can be seen more clearly than in the painted counterproof of Plate 15. In the unpainted etching the transparent nature of the cocoon is indicated by a series of curved lines running horizontally (Figure 71). These fine lines are largely masked in the painted counterproof.

FIGURE 71   Merian, 1679. *Raupen*. Detail from Plate 15 showing eggs
             and the transparent nature of the cocoon layered over
             the pupa of *Spiris striata* (feathered footman)

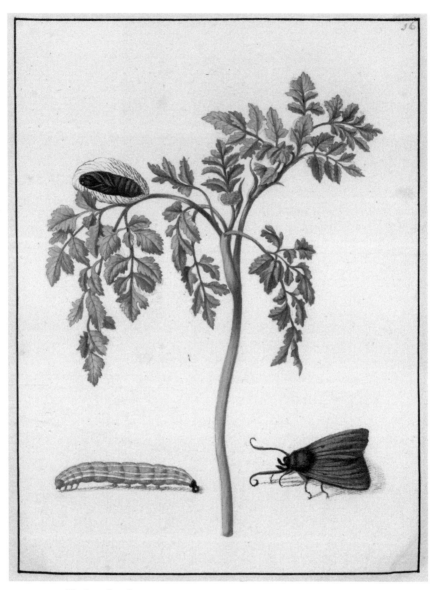

PLATE 16   Merian, 1679. *Raupen*

xvi. Chervil   *Cerefolium*

(33) It is almost unbelievable how many different caterpillars tend to a green color, so that at first some are readily distinguishable from the others, some just barely, and some not at all. The size, fatness, and shape, as well as their color and markings—in the form of lines, blotches, spots, and such—are often so exactly alike that even the most discerning observer cannot easily distinguish them unless he examines them either with the *microscopium*, as the magnifying glass is called, or by capturing, feeding, and maintaining the caterpillars. Then he can observe and conclude from their food; or their initial, intermediate, and final stages of change; or, finally, from the correct, natural [adult] insects that come forth (whereas some undergo a false change into small flies and such) what noticeable distinctions do or do not exist among them. All of this I have experienced sufficiently with many kinds of green caterpillars. But this caterpillar crawling at the bottom, which could be clearly recognized among the others and distinguished from them, appeared attractively striped green and white. Everything it gave off [excreted] was quite black, which is not the case with other caterpillars. It had six little green claws, (34) eight round, green legs in the middle of its body, and two of the same color at its hind end. Such ones I have found in early May on chervil and fed them on it until the end of that month. Then it began to change and made a thin, white cocoon around itself, in which it became a chestnut-brown date seed resembling a swaddled baby, like the one resting on the arched stem of chervil. After fourteen days, a moth-bird came out that was all brown and speckled with darker brown spots, having six legs and two larger and two smaller horns. The caterpillar moved very slowly, and the moth-bird flew only at night, when it shone very nicely. It is pictured resting at the bottom, next to the green caterpillar.

Chapter 16: *Amphipyra livida* (noctuid moth) and *Anthriscus cerefolium* (chervil)

No common name has been given to this rather plain looking noctuid moth that arises from a bright green caterpillar. Merian writes that the moth "shone very nicely" when flying at night. This observation seems inconsistent with the very dark color of the top of the moth, but the undersides of its wings are a light iridescent color, and they would indeed reflect the light from any nearby source (Figure 72). She raised its caterpillar on chervil, pictured in plate 16, but the species eats other herbaceous plants as well.

The information specific to this moth is congruent with the correlated handwritten entry in the study journal, but as elsewhere in the *Raupen* book, she inserts some of her broader conclusions regarding lepidopterans. These generalizations – such as the traits differentiating moths and butterflies – were the result of months and years of accumulated observations and were likely formulated after she recorded her initial notes on each species. Here she takes the opportunity to remark upon the large number of green caterpillars that transform into very different adults, and she includes the statement that it requires the use of a magnifying glass to differentiate among similar larvae. Alternatively, she offers the suggestion that one could capture and raise caterpillars to adults, with the added benefit of learning about their food, their stages of change, and in some cases, about the false changes that we now understand as emergence of parasitic insects. The knowledge Merian gained by raising dozens of species was supplemented by what she observed using magnification.

FIGURE 72   Photograph of underside of *Amphipyra livida* showing iridescence of the lower
wings. Dumi, Creative Commons

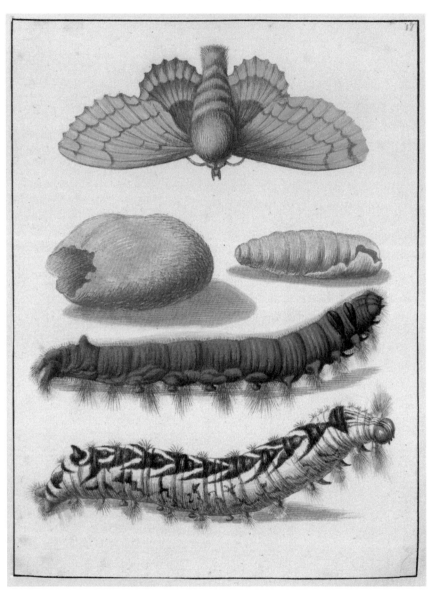

PLATE 17   Merian, 1679. *Raupen*

XVII. Miraculous caterpillars *Vermes miraculosi*

(35) These two large caterpillars, so unlike each other in their form and color, appeared at the same time, namely in May. They were several times provided to me by distinguished amateur collectors [*Liebhaber*], who sent word and also told me themselves that they had discovered them in the grass. Accordingly, I cannot attest to their having any other food than this, although I readily believe their report because the caterpillar appearing in Plate XXXII likewise ate grass under my observation; and all three of them turned out exactly alike, whether in spinning their cocoons, in their date seeds, or as adult moth-birds, so that no difference could be detected among them. The lower large caterpillar is white, with large black markings on its back and toward both sides, and somewhat lighter, smaller ones lower down. At the front of its head it has two brown tufts of hair, and on all its other segments, lower on each side, long, round, wart-like protuberances with brown hairs. On the top of its hind end is a blunt horn. As for the rest: six large claws, eight large legs, and at the rear two more of an odd kind. On the evening of the day when I received it, it entered its (36) change, and within several hours had spun itself up in nothing but hairs, so that I had to paint it from midnight on until morning, since I wanted to capture and present its marbled appearance; for at the break of day it was entirely spun up and looked gray, like the oblong cocoon lying above the upper caterpillar. Inside, it had pushed off its skin and turned into a gray date seed like the one lying next to it, and so it remained into June, when a large moth-bird came out that was dark orange in color, had two black horns and six black legs, which were brown at the tips. It flies only at night, creates a great fluttering, and moves very nimbly.

The other large caterpillar, however, is dark brown, not so fat, but shapelier; it has two black patches like velvet on its head, only five (but longer) warts or bumps in a line between the head and the first round leg. At the rear and on top is a round, blunt horn, and below, brown hairs sprout everywhere. This one also makes a gray cocoon as described above, and inside it the same kind of date seed, after pushing off its skin and before the insect comes out. Both of them lay white eggs marked with round, sea-green spots. Now, where such dissimilar caterpillars come from, even though [they produce] the same kind of insect, the naturalists can perhaps explain better.

Chapter 17: *Gastropacha quercifolia* male (lappet moth)

The lappet moth caterpillars in Plate 17 are the only ones pictured in her 1679 *Raupen* book not collected by Merian, but that were instead provided by "distinguished amateurs." She reports here that she received caterpillars on multiple occasions from *Liebhaber*[1] who found them in the grass. Lappet moth larvae typically eat a variety of woody plants, including willow and various fruit trees. The large male lappet moth is not shown with a plant, perhaps because by depicting it and the caterpillars as life-size, there was little space for any foliage. Alternatively, she may have omitted a plant because she was unsure of the food preferred by the caterpillar. However, in Plate 32, to which she refers in the text here, she pictures a female adult lappet moth and shows the caterpillar on grass, stating that this is where she found them. Like the helpful "amateurs" who gave her the caterpillars pictured in Plate 17, she probably collected these late instar larvae a short time before pupation, when they stop feeding and move to the ground to seek a spot to transform. It is easy to find the caterpillars at this stage, and as they no longer eat at this point, having the food plant for them would not be essential to obtaining adults from caterpillars.

Plates 17 and 32 provide a third instance in the 1679 *Raupen* book in which sexual dimorphism of the adults has led Merian to portray male and female in separate chapters. In this case, she makes it clear that she was aware that the adults portrayed represented the two sexes of the same type of moth. She cross-references these entries in both chapters, making note of the similarities of the adult moths, and she uses this opportunity to show one moth as it would appear when flying and the other in a resting pose. Indeed, Plates 17 and 32 are rare examples of compositions in which Merian shows a moth with wings in two different positions. The female moth in the latter plate is shown in profile and in the act of laying her eggs, as opposed to the male in the former, pictured with his large wings spread across the page.

The most intriguing discovery made by Merian regarding the lappet moths was that the larvae are highly variable in coloration (compare the three larvae pictured in Plates 17 and 32). She was so amazed that such different-looking larvae would lead to identical pupae and ultimately the same type of adult that she termed them *Vermes miraculosi* (miracle worms). One caterpillar began

---

1　*Liefhebbers* (Dutch) or *Liebhaber* (German) are useful terms that have no precise modern equivalent, but that mean something akin to "amateurs of nature and art." Many of these "amateurs" were very dedicated to their pastimes, often making substantial contributions in their area of concentration.

changing the day she received it, causing her to stay up all night painting it before it completed formation of its pupa. In Chapter 32, she describes a third color pattern and some further structural details of the lappet caterpillars.

Merian appears to be the first naturalist to have published accounts of the striking polymorphisms among the larvae of a given insect species. She was puzzled as to how these dissimilar caterpillars gave rise to the same adults, but she reported what she saw, and as with the mysterious appearance of flies and wasps from the bodies of caterpillars, suggested that perhaps other naturalists could explain the phenomenon. Such larval polymorphisms have been a question of interest for centuries. In 1893, Edward Poulton raised *Gastropacha quercifolia* caterpillars on a variety of branches covered with differently colored lichens. He found that after overwintering, the caterpillars matched their backgrounds very closely. Contemporary researchers have explored a variety of causal mechanisms in larval color variations ranging from diet, to light, temperature, and hydration status of the larvae, and continue to study the phenomenon centuries after Merian noted these polymorphisms.[2]

---

2   Poulton, Edward B. 1903. Experiments in 1893, 1894, and 1896 upon the colour-relation between lepidopterous larvæ and their surroundings, and especially the effect of lichen-covered bark upon *Odontopera bidentata, Gastropacha quercifolia*, etc. *Transactions of the Royal Entomological Society of London, 51*(3): 311–74. For a more recent study, see Sandre, Siiri-Li, Ants Kaasik, Ute Eulitz, and Toomas Tammaru. 2013. Phenotypic plasticity in a generalist insect herbivore with the combined use of direct and indirect cues. *Oikos, 122*(11): 1626–35.

PLATE 18   Merian, 1679. *Raupen*

XVIII. Honey apples, in flower *Malus mellea, florens*

(37) This type of caterpillar feeds on all kinds of trees and is very partial to moisture and drinking. Thus, in this year of 1679, with its weather of alternating temperate rains and bright sunshine throughout the spring, they numbered many thousands and were of exceptional size, since they were regularly refreshed and grew steadily larger. For when the rainfall is too frequent, heavy, and sustained, it lacerates the young caterpillars, and the ones that seek shelter and safety on the undersides of green leaves are washed away by it and are destroyed in the runoff; the same happens to the eggs. Or if there is too long a period of fair weather, many of them also die, especially if they have not yet spun cocoons or been able to find other sources of moisture.

They have done enormous damage not only to fruit trees but also to lindens and willows—in fact, all kinds of other trees. Their trunks were entirely covered from top to bottom and their leaves often stripped so bare that next to other trees they looked as if frost-killed. But this will serve as useful information for attentive gardeners: namely, as evening approaches, all of them crawl together onto one branch where they have their cocoon or nest and (38) can be captured all at once and destroyed.

The caterpillar resting on the apple blossom and two green leaves appears gray with a yellow head and black stripes and blotches. On their first five segments they have ten raised, round, black grains; and on each hindmost segment two shiny, ruby-like dots from which stiff, black hairs grow. They have six claws, eight legs in the middle and two more behind, all of them brownish in color. They change, some in June and some in July, and usually turn directly into a brown date seed like the one depicted at the bottom, opposite the adult insect. Some, however, spin a light cocoon and lie dormant in that form for fourteen days. Then a large moth comes forth, white with brown markings. It has two brown eyes and horns and six legs of the same color. By day, they appear to be dead, can be tossed around and remain inert; but at night they fly about quite merrily. After several days they lay their eggs, as can be seen next to the moth. Tiny, entirely black caterpillars crawl out in April of the following year. They make a kind of coat around the eggs such that one will not even notice them, and they are thus well protected from cold—as the winter just past was very cold, yet they were not harmed by it.

Chapter 18: *Lymantria dispar* (female gypsy moth) and *Malus domestica* (apple)

In her account of the voracious gypsy moth, Merian addresses in more detail the 1679 population explosion of caterpillars first mentioned in Chapter 10, and she attributes the increase in numbers to the powerful effects of weather on her insect study subjects. She states that warm rain followed by sun contributed to a large number of these moisture-loving larvae. However, she adds that heavy rain is detrimental to these insects, lacerating the soft-bodied caterpillars and washing away eggs, and notes that caterpillars will shelter under leaves in rain. The corresponding study journal entry on the gypsy moth is longer than for most of her subjects, and in these research notes, Merian includes her observations on the effects of weather on the caterpillars.[1]

The consequence of the gypsy moth population spike was herbivorous mutilation of all sorts of trees, which looked as if they were frost-damaged. Merian clearly meant for her book to be useful to gardeners, and thus makes a point to mention the communal "nests" or webs that the caterpillars occupy in the evening; she then recommends that gardeners use this information to collect the destructive caterpillars *en masse* for eradication. In addition to the usual description of pupation and the physical characteristics of larva, pupa and adult, she discusses the gypsy moth eggs in unusual detail. As she correctly describes, the species overwinters as eggs that are protected by a coating provided by the female.[2] Merian marveled at the protection this provided against an exceptionally cold winter.

Later in the book, Chapter 31 appears to depict a male gypsy moth, and again she recounts its damaging habits, this time on roses, fruit and other types of trees. As with other males, she did not include eggs in that plate. In this plate, like Plate 1 of the silkworm moths and Plates 32, 37, and 46, Merian depicts the female depositing eggs, something not illustrated in Aldrovandi's, Moffet's or Jonston's books on insects.[3]

---

1   Appendix, study journal entry 38.
2   The covering is made of setae or "hairs" from the abdomen of the female. Eggs are also protected by their location in crevices and by snow cover, and they can survive temperatures below −32 °C. Leonard, David. E. 1972. Survival in a gypsy moth population exposed to low winter temperatures. *Environmental Entomology*, 1(5): 549–54.
3   Moffet's volume dedicated exclusively to silkworms has only one plate, and this shows one adult moth, one caterpillar, and the cocoon-covered pupa. Moffet, Thomas. 1599. *The Silkewormes and their Flies*. London: By Valentine Simmes for Nicholas Ling.

PLATE 19   Merian, 1679. *Raupen*

XIX. Flesh-colored rose　　　　*Rosa incarnata*

(39) Of all the flowers, herbs, and trees, it would be difficult to find any one kind that has not had, and does not still have, one or more caterpillars as its natural enemies, since many are sure to be found feeding on one flower at the same time, as this flesh-colored rose and various other roses to follow can thoroughly attest. For the very small caterpillars attack not only its green leaves, but also eat away the hearts of the unfolding rosebuds themselves so completely that a rose can no longer develop out of them; and this they do until the end of May. Now these caterpillars are for the most part a pretty green. Their head, together with the six claw-like forelegs and their hind segments, is black. In the middle of their body they have four small legs on each side. One such caterpillar I have placed at the top, on the rosebud. If one should touch these caterpillars, they will immediately let themselves down to the ground by a thread and climb back up on it. They are also very nimble and quick in moving about. Concerning their food, I have maintained them on rosebuds up to the end of May; then they went (40) into their change, and turned into date seeds (one of these is hanging from a stem opposite the bud), in which form they remained until the twelfth of June. At that time, small brown moth-birds, or *Schaben*, came out. These had two brown horns and four legs. One is pictured at the top, resting on the rose.

The other caterpillar, which is liver-colored and quite slender, I have pictured at the very bottom, on a leaf branching from the stem. With this one it was the same as with the previous one, in that I also maintained it with rosebuds. Behind its head it has three claws on each side, and in the middle of its body four legs, again counting those on each side, and at the hind end two more legs, all of them of that same liver color as the caterpillar itself. At the end of May, it entered into its change and became a wood-yellow date seed, from which after twelve days a charming and nimble moth-bird emerged. It had two wood-yellow horns, its head the same color, as were the forewings. Its body and legs were also that color and streaked, but the two hind wings were a silver color. The flying insect at the bottom of the plate and its empty date seed hanging not far away on a badly chewed leaf show this satisfactorily.

Chapter 19: Tortricid moths and *Rosa* sp. (rose)

The two small moths shown with the rose are in the Tortricid family, but it is not possible to identify them reliably to species based upon Merian's image. Of all caterpillar behaviors, Merian seems most charmed by that in which the larvae lower and raise themselves from vegetation by means of a "thread" from their mouths. This chapter provides her second mention of this phenomenon, but she does not illustrate it in the image. The plate shows a rose stem exhibiting extensive insect damage to the leaves, and part of this composition is copied from the rose image in Plate 11 of her first *Blumen* book (Figure 42 in Chapter 2).[1] Interestingly, Merian includes only one rose among the three dozen plates of flowers in her *Blumen* books, and of all the flowers included in these images, the rose shows the most leaf damage due to herbivory. Another ornamental rose is woven into the wreath in the frontispiece of the third *Blumen* book (published in 1680), and again, it is the only flower on that plate with an insect-damaged leaf.

Whereas Merian usually fed her captive larvae on the foliage of their host plants, she raised the Tortricid caterpillars in Plate 19 to pupation on buds from roses. Her text introduces the fact that caterpillars of these two types of moths will eat the buds to the point that no flower can form. In Chapter 24, she illustrates another Tortricid caterpillar eating its way out of a rose bud, and again she raised these on buds rather than foliage. Although such bud-eating pests are the bane of rose gardeners, this type of herbivory is even more devastating in a natural habitat. In a wild plant that is not cultivated by cuttings as roses are, bud damage can effectively neuter the plant and severely decrease the plant's population.

Merian also showed insects on roses in Chapters 22, 24, and 28, possibly as an aid to marketing her book to gardeners, many of whom would be likely to grow these ornamental flowers. Early in her study journal, she commented on three types of rose-eating caterpillars that she pictures in the small study on vellum;[2] one is the moth seen here in the lower portion of Plate 19, and the other two are featured in Chapters 24 and 28. Her notes on the moth in the upper portion of Plate 19 appear separately in study journal entry 35. Hence, it is possible that she combined into one plate two species of insects that she

---

1   Merian, 1675–1680 *Florum fasciculus Primus* [– *Tertius*]. Leaf damage to a lesser degree appears in Plates 2, 5 and 8 of the 1675 volume. No leaf damage is shown in the second volume of 12 plates (1677). In the third volume (1680), a narcissus shows damage in Plate 7, and there is one tiny hole in a leaf in Plate 10.
2   Appendix, study journal entry 4.

studied at different times, possibly in different years. Unfortunately, there are few dates in her study journal and, as mentioned earlier, a clear chronology for these research notes remains elusive.

Merian included some details in the Tortricid journal entry that were omitted in her published text. In the journal she writes that "one sees them crawling through the hollowed-out stem" and speculates that the caterpillars grow within the plant. She also remarks that the three types of caterpillars shown in study journal entry 4 "take all kinds of roses as their food and do much damage to them." In Chapter 19 she introduces her conclusion that all plants seem to suffer caterpillar damage, and she adds that roses may be attacked by several types of insect at a time.[3] Both Chapters 18 and 19 make strong arguments for the importance of understanding the food plants of insects, a case that she builds throughout this and her later books.

---

3   A number of fruit species are in the rose family (Rosaceae), including cherries, apples, plums and more, and these are similarly afflicted by insect pests; however, many plants have multiple insect pests.

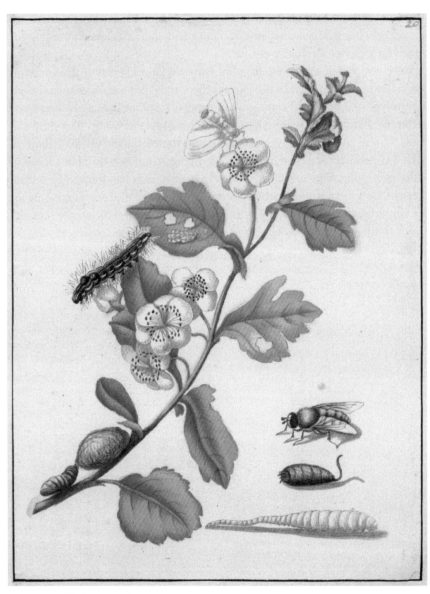

PLATE 20   Merian, 1679. *Raupen*

xx. Hawthorn in flower *Oxyacantha florens*

(41) Caterpillars of this type I have found in May, mostly on the hedges where there was much hawthorn or *Hagbutzen*, which they used as their food. But if it happens that they find none of those leaves, they will eat other things as well. Thus, I have given them quince, rose, hazelnut, and wallflower leaves, all of which they devoured. Now this type, one of which is crawling over the little cluster of the hawthorn's blossoms, is bright red with black and white stripes, its head black and dark-striped. It is also covered with little black hairs. Behind the head it has six black claws on both sides, eight legs in the middle of its body and two at the back, and all of them appearing black. At the end of May it went into its change, making a gray cocoon in which it turned into a brown date seed that was very restless when touched. But after twelve days, a moth-bird like the one resting on the upper blossom came forth. It was a pretty white and had, at the rear of its body, something like a small yellow (42) fur pelt from which it discharged its light-yellow eggs immediately after [it emerged from pupation]. They can be seen behind the caterpillar on an eaten-away leaf. It also has four white legs and two white horns on its head; it does not fly high.

As an additional offering, farther down in this plate, an amusing white worm of which I have found many at home, in old holes in the walls and in the cellar, in dirty sand where they fed and underwent their change; [their pupa is] like a tiny mouse in form, mouse-colored as well, and appearing speckled. For their change they took seventeen days. After that, flies resembling bees crawled out, yellow and brown in color, on which at first no wings could be detected, but only something small and folded together on both sides, which they pulled open with the hindmost of their six legs so that finally wings appeared; and all in the space of one hour. One finds many of them on all sorts of field flowers, from which they suck the honey- or sugar-juice. So I have also given them sugar, which they readily ate. They are very slow, lazy, and lethargic in walking as well as in flying, and I have shown one above the changed form, or date seed, for the sake of a better report.

Chapter 20: *Euproctis similis* (yellow tail moth), *Eristalis* sp. (hoverfly or 'rat-tail maggot') and *Crataegus monogyna* (common hawthorn)

The hoverfly, also known as a dronefly, is unusual in the *Raupen* book in that it is has no connection with either the moth or the plant figured with it in the plate. This unusual insect is presented as an "additional offering," possibly to fill a large area below the main figures. The adult does resemble a bee and feeds on the nectar of flowers, as she writes. Her language mirrors that of Goedaert's, who depicted this insect in his second plate. Goedaert wrote, "the nourishment of this bee is a certain sweet juice that it draws from the flowers."[1] The "amusing" larva of the hoverfly is aptly called a rat-tailed maggot. In a late 19th century investigation of these animals, George Buckton mentions Réaumur as being the first to describe the insect, although this clearly was not the case.[2]

The small but elegant yellow tail moth is another species described by Merian as laying its eggs immediately after emerging from pupation. The pupa is depicted balanced precariously on the stem of the hawthorn, just below the cocoon, and seeming to touch it. In reality, the cocoon would cover the pupa, and if the change from the caterpillar had not occurred in captivity, pupation would have occurred on the ground and the cocoon would have been hidden among dead leaves. Either because she did not observe any of these cocoons in the field or because this would be difficult to depict, she used the space-saving but somewhat awkward composition seen in Plate 20. The moth is shown in the act of landing on a flower, and she notes that these moths do not fly very high, an observation she must have made in the field. Once again, she mentions four legs in the adult, because this is what one sees when the forelegs are folded against the body. As she asserts, the larvae of the yellow tail moth are generalists, devouring a wide variety of plants.

1   Goedaert, *Metamorphosis*, Volume 2: 24.
2   Buckton, George. 1895. *Natural History of Eristalis tenax or the Drone-fly.* London and New York: Macmillan and Company: 4.

PLATE 21   Merian, 1679. *Raupen*

XXI. Quince blossom   *Cotoneæ flos*

(43) All the caterpillars of this large type have an orange-colored head with black spots and, in the middle, a white stripe; they are wood-yellow in color. Between their segments they have wide, black bands running down, which appear like the blackest velvet and on which they also have small, snow-white spots. The hindmost segment, however, is of the same color as the head. A not badly fashioned example of this caterpillar can be seen between the green leaves of the attractively colored quince, the same on which I fed and nourished it into the month of July. But it was also very inclined to drink, for if I had not constantly given it water it would have died, as has happened to them and to other caterpillars of similar habit in my keeping. It also had this peculiar way, when only slightly touched, of thrashing about violently with its head as if it were angry and could not endure being touched. So after it began its change, in that same month, and spun a wood-yellow, oblong, and (44) very hard cocoon (one of which lies completely opened in the lower corner), it remained in that cocoon into the month of August. Then large, fat moth-birds came out, also of a wood-like color, but they did not move all day long. However, as soon as they were touched, they would fly headlong into the nearest wall and fall to the ground. Thus, I am convinced that they cannot see well by day, for which reason one does not notice them flying very much, any more than other moth-birds, in the middle of the day, but only toward evening or at night (as I have already mentioned a number of times in the descriptions of other caterpillars), unless they have been disturbed or stirred up by someone. After several days, they laid their eggs, which have a color like the moth-bird's and are shown on the green leaf in the center foreground. The attractively speckled insect in flight at the top is easily recognized as well.

Chapter 21: *Lasiocampa quercus* female (oak eggar) and *Cydonia oblonga* (quince)

The full life cycle of the female oak eggar from egg to adult is shown on the quince tree that Merian used to feed the larvae. Quince is native to southwest Asia and had not been long established in Germany when she pictured it in her book, so this inclusion may have been something meant to interest a gardening audience. The caterpillar has left behind a damaged leaf and seems poised to take a bite from a new area of the plant.

When compared to the male of the species pictured in plate 10, the sexual dimorphism in color, antennae, and the adults' abdominal size and shape is clear. The contrast is easy to make, and her use of the view of both adults with wings spread may have been intended for this purpose. She again emphasizes the need these caterpillars have for water and their habit of drinking, and she adds that if one is touched, it thrashes its head about "as if it were angry."

The pupation of this species is described as lasting from July into August in this chapter, and from June into July in the earlier entry regarding the male. It may be that she raised these adults in different years, and that weather variables shifted the timing of their metamorphosis. This timing difference, and the variation in the caterpillars' coloring, may be two reasons why she did not explain that she was showing the same type of insect in Plates 10 and 21.

Merian speculates in this chapter and elsewhere (e.g., Chapters 31 and 47) that moths fly by night because they cannot see well by day. In fact, night-flying moths can sense ultraviolet radiation and polarized light, but what she says is in essence true, and night-flying moths rely more on scent (hence the more elaborate antennae) than on vision to find mates; the converse is true for diurnal (day-flying) moths and butterflies. She implies that butterflies do see well by day, and in her preface refers to them as having "bright eyes."

PLATE 22   Merian, 1679. *Raupen*

XXII. Small, hundred-petal rose

Rosa multiplex, media

(45) Here we have another of that kind of caterpillar which, like the ones mentioned earlier, takes roses as its food, eating them away, the green leaves included, but especially the young, green buds, hollowing them out completely and ruining them. Now these caterpillars are yellow and have a red stripe from their head halfway down their body, and they spin with their mouth, as can be seen on the lower rosebud. However, they change in two different manners. In one, they lie down and turn into a date seed, like the one shown lying on the arched stem; it is pink and green and remains lying motionless into July. Then a most charming, beautiful insect emerges which looks entirely different from the previous moths or summer-birds. It is white and wood-yellow and has a very rapid flight. For when it flies off, one has great difficulty to capture it clean and intact. But some of this type of caterpillar come to rest and appear dead, and in several days, a number of small maggots crawl out of them and immediately (46) spin into little white capsules. But the old mother caterpillar spins the just-created five little date seeds all together, and after that, she dies. Then, in fourteen days, a little fly comes out of each small capsule. Two are seen flying and two sitting on the leaf with holes in it, also the little caterpillars and capsules spun up together and lying nearby. What the true cause of such disorderly changes might be, namely that these two dissimilar little creatures came from a single kind of caterpillar, whether perhaps the result of their imperfection or something malignant about them, I have not been able to discover or to imagine, but have had to leave to learned men. I must note only this in passing: What a diligent and artful love Nature itself has implanted in such insignificant, negligible, and yet (even though destructive) dainty little creatures; namely, that the mother caterpillar, because her life cannot last any longer after her false egg-laying, seeks first to spin her young together and thus to protect them from any harm, so that none of them might be separated from the others or perish. To which end, as we have said, before her (so to speak) maternal farewell, she had to unite her young and connect them to each other most carefully by this natural bond.

Chapter 22: *Cnaemidophorus rhododactyla* (plume moth) and
Rosa sp. (rose) with Braconid parasite

The plume moth is another whose larvae feed heavily on roses, especially the bud. This insect is unique in the *Raupen* book in belonging to the Pterophoridae family, known for their feathery wings. Merian notes the singular appearance of this "charming, beautiful" moth, but spends most of the entry recounting the two different transformations she witnessed: one a typical metamorphosis of the caterpillar through pupation into an adult moth, and the other in which small maggots emerge from the body of the caterpillar and then are "spun up together." She mistakenly attributes the formation of the cocoon around the pupae of the parasitic wasp (a Braconid) to the "old mother caterpillar,"[1] but she is puzzled by the "disorderly change" that results in what she calls little flies. It made no sense to Merian that two kinds of adult insect developed from one type of caterpillar, a mystery that she understood by the publication of her final caterpillar book, but the solution of the present case she leaves to "learned men."

She ends her speculation on this peculiar phenomenon with a tribute to Nature for the "attentive and artful love" demonstrated in the protective maternal care of the dying "mother caterpillar." Her wording in this entry echoes that of Goedaert in his description of the same type of occurrence in the caterpillar of the cabbage white (featured in Merian's Chapter 45). He wrote that the flies came from the "mother caterpillar" that "spun all of the small silk houses together firmly, as if with a bond of love" so that they would find each other after their transformation.[2] Merian did not include any speculation on the cause of the disorderly change in her study journal, and it may be that she searched Goedaert and perhaps other sources for answers to this mystery.

1 Some Braconid larvae spin small individual cocoons, which, when they are in a mass on a dying caterpillar, appear to be one large cocoon encasing all of them, as pictured by Merian. Larvae of wasps from the genus *Apanteles* form such masses, and several of these species lay eggs in the plume moth. Balevski, Nikolai A. 1999. *Catalogue of the Braconid parasitoids (Hymenoptera: Braconidae) Isolated from Various Phytophagous Insect Hosts in Bulgaria.* Sofia: Pensoft: 48–50.
2 Goedaert, *Metamorphosis Naturalis*, Volume 1: 43.

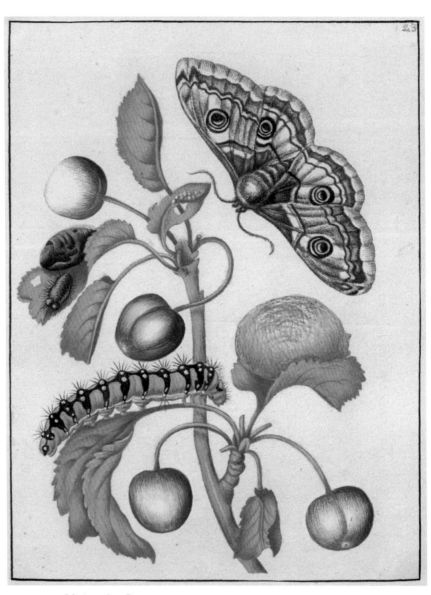

PLATE 23   Merian, 1679. *Raupen*

XXIII. Sweet, early cherries

Cerasus major, fructu subdulci

(47) Many years ago, when I first saw this moth, large and exceptionally well formed by Nature, I could not marvel enough at its beautiful shadings and contrasting colors, and at the time I often used it in my paintings. But after I discovered the transformation of the caterpillar some years later, by the grace of God, it seemed to me a very long time until the beautiful moth-bird came forth. Thus, when I did obtain it, I was filled with such great joy and was so pleased in my intent that I can hardly describe it. But over time, I have had the caterpillars for several years in succession, feeding them on the leaves of sweet cherry, apple, pear, and plum into July. They had such a lovely green color, like young grass in the spring, with a beautiful, straight, black stripe along the length of their back; and running downward on each segment, a black band on which four little round, white grains like pearls shone. Below them is an elongated golden-yellow spot, and below that another white pearl. Beneath the first three segments, (48) on both sides they have three red claws, then two free segments, and behind those, four green legs the same shade as the rest of the caterpillar, and one more on each side of the last segment. On each of the pearls just described, a long black hair sprouts along with other smaller ones, and they are almost stiff enough that one could be pricked by them. It is curious that this type of caterpillars, when they have no food, will devour each other out of sheer hunger, but will cease doing so as soon as they have food again. Now when such a caterpillar has attained its full size, as seen here on a green leaf and its stem, it makes a hard cocoon, shiny as silver and oval-round, in which it first pushes off its very small, rolled-up skin and changes into a liver-colored date seed, which is seen, together with the shed skin, lying above the caterpillar. It remains thus unmoving until mid-August, when at last the lovely moth-bird extolled above and shown here in flight comes forth. It is white, has gray-flecked patches, two yellow eyes, and two brown horns. On each of its four wings is a pattern of circles within and around each other, which are black and white and yellow; on the edges of the wings it is brown, but near the tips (which is to say, only on one side of the two outermost wing tips) it also has two beautiful rose-colored spots. By day it is quiet, but at night very restless.

Chapter 23: *Saturnia pavonia* female (emperor moth) and *Prunus avium* (wild cherry)

The female emperor moth shown in Plate 23 is strikingly different in color and size from the smaller, brighter male depicted in Plate 13, so perhaps it is not surprising that Merian treats them separately. Here she includes the eggs in her image, but does not describe them in her text. In this composition, she shows the eggs on a leaf, whereas this moth usually lays them encircling a stem or twig; it may be that in captivity, the eggs were not deposited in the usual way, and she was unable to observe the natural pattern of oviposition.

In the entries on both sexes of the emperor moth, she lists various fruit trees as the larval food, but expands the entry here to recount the cannibalism that occurs in caterpillars deprived of food, something mentioned again in Chapter 35 regarding the dot moth. She goes on to say that both the emperor moth and dot moth caterpillars return to their plant food when it is available, so the cannibalism she observed may have been an artifact of captivity. However, a number of lepidopterans are partially carnivorous, and many exhibit cannibalism, particularly when water-stressed or nutrient-deprived.[1] She was not the first to note this behavior; Goedaert writes about caterpillars eating one another, and cites one type (not identifiable) that does this "greedily."[2] Merian appears to have been unfazed by this rather dramatic behavior, and spends much of this entry expressing her delight in the beauty of the adult and her pleasure in obtaining it after much effort.

She also extols the "lovely green" of the caterpillar, like that of young spring grass, and she describes and illustrates the black markings of these larvae. However, the caterpillar depicted in Merian's study journal entry for the female emperor moth has no black markings and is instead the green morph that is also common in this species. Once again, she made the decision to edit what she shows in her final plate. It appears that for Plate 23 of the female emperor moth she reused the caterpillar image from Plate 13 (see Figure 69), providing us with further hints that she was aware that these were the same type of adult moth. She employs views of both male and female moths with wings spread, making comparison of the two easy for her readers.

---

1   Reviewed in Pierce, Naomi. E. 1995. Predatory and parasitic Lepidoptera: carnivores living on plants. *Journal of the Lepidopterists Society*, 49(4): 412–453.

2   Goedaert, *Metamorphosis Naturalis*, Volume 2, Plate 32. It is often difficult to identify his insects to species.

PLATE 24   Merian, 1679. *Raupen*

XXIV. Large, hundred-petal rose

Rosa maxima, multiplex

(49) Certain things have often struck me as curious (over the course of my investigations), and so I thought these little caterpillars were unusual, for they not only take all kinds of roses for their food and do great damage to them, but also even lodge themselves in the hearts of the young buds and eat away at them, so that one often does not see them at first unless one breaks open the bud, not knowing how they came to be there or from what they grew. I have pictured the head of one of them crawling out of the lowermost rosebud, having eaten its way through (something I have often observed). The entire caterpillar is seen above it, resting on the middle rosebud. They are a pleasant green and have a black head, also three black claws on either side, and legs like those of most other caterpillars. When touched, they let themselves down to the ground on a thread, by which they climb back up to where they had been. I maintained them on rosebuds until the end of May; at that time they changed into a brown date seed, one of which is resting on the stem of (50) green leaves. They remained resting and undergoing this change until mid-June, when very pretty moth-birds came forth, which likewise tended to stay near the roses. One can be seen in flight above, and one resting on the rose at the top. These moth-birds are not only attractive but also appropriately named, for they are exceedingly agile and quick in their flight. They have small brown eyes, two yellow horns, and six yellow legs. The head and the forewings are a pretty speckled yellow and have a sheen as if they were of gold; the two hind or lower wings, by contrast, are gray and like silver. Furthermore, it is surprising that this insect is so small and yet has such unusual characteristics: namely, that it flies by day, whereas other moths fly in the evening or at night. And if one wishes to catch it, it is so quick to conceal itself underneath the green leaves that one does not see where it has gone. Thus, it is not easy to capture unless one obtains it by raising the caterpillar.

Chapter 24: *Acleris bergmanniana* (yellow rose button) and
Rosa sp. (rose)

The yellow rose button moth is the fourth in the family Tortricidae figured by
Merian, and as with the two Tortricids shown in Plate 19, she once again emphasizes the damage these caterpillars inflict on roses. She was intrigued by
their habit of living inside the rose buds and eating them from within, and
in addition to the usual depiction of the entire caterpillar, she added a second one crawling out of a bud (Figure 73). Like related species of caterpillars,
these use lifelines spun of silk to escape when touched. Although she does not
mention it, the yellow rose button is also a leaf roller like the Tortricid moth
described in Chapter 7. Chapter 24 is one of the few entries that mention the
escape behavior of the adult moth, which she says hides under leaves when
pursued, remarking that it is easier to get the adult by raising it from a caterpillar than to capture it. She also notes that this moth has the unusual habit of
flying by day, which is the case for other moths in this family.[1]

The rose pictured in this plate is interesting in its own right, and gives us a
clue as to how Merian reused images. She painted a very similar rose at least
two more times in the friendship albums of Christoph Arnold and his son
Andreas Arnold. In the album paintings, the plant is simplified, presenting two
fewer buds and three leaves rather than five.[2]

---

1 Tortricid moths often fly by day. Today, researchers can determine patterns of activity by
   setting traps for the insects and by using motion detectors, but Merian simply made field
   observations. Knight, Alan L., Weiss, Mark, and Tom Weissling. 1994. Knight, Alan L., Mark
   Weiss, and Tom Weissling. 1994. Diurnal patterns of adult activity of four orchard pests
   (Lepidoptera: Tortricidae) measured by timing trap and actograph. *J. Agric. Entomology*, 11(2):
   125–36.
2 Taegert, Werner. 1998. Man's life is like a flower – Maria Sibylla Merian's watercolors for *Alba
   Amicorum*. In Merian and Wettengl. 1998. *Maria Sibylla Merian*: 89.

FIGURE 73
Detail from plate 4 of Merian's 1679
Raupen. Acleris bergmanniana
(yellow rose button)

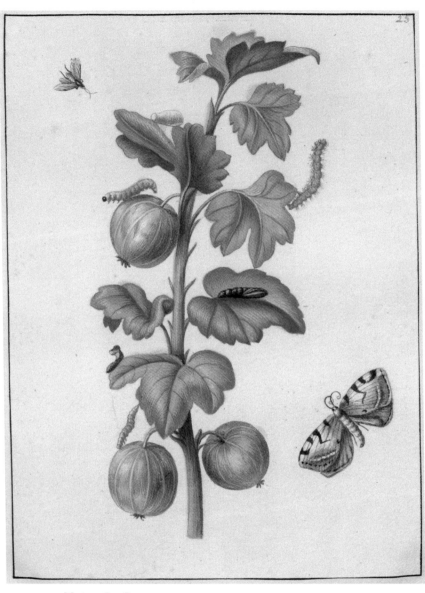

PLATE 25   Merian, 1679. *Raupen*

XXV. Large Spanish gooseberry

Fructus grossulari sativæ, spinosæ

(51) That there are countless caterpillars, large and small, on gooseberries as well as on roses is unfortunately all too true for many a gardener, considering that they can inflict great damage on both plants. So it is that I found these two caterpillars on the gooseberries, also called *Klosterbeeren*. Both of them take these as their food. If one touches the first kind, like the caterpillar standing upright on one green leaf, it holds fast with its hind legs and thrashes about violently with the front part of its body as if it were angry; in fact, it fastens itself so tightly that one can hardly pluck it off without injuring it. When it moves, it draws its hind legs up to its front ones so that its body arches up high. It is a pretty green in color, speckled with black spots, and has a yellow stripe below. From the head back, six greenish and somewhat darker-shaded claws, just behind them, two, and at the hind end another two legs, as well as tiny green hairs. After it had fattened itself well on the above-mentioned leaves and the time of its changing had come, (52) at the end of May, it changed into the date seed, which was brown, pictured on the leaf below the caterpillar. In the middle of the following month, June, the insect seen flying upward from below emerged from it. It is bright brown, but with brown patches above, and small black spots all over. It flies very swiftly at night.

The other type of caterpillar is also very often found in May on the gooseberry fruit and leaves. They eat them away so completely that nothing but the stem remains. The caterpillar itself is green, but toward its black head and at its hind end somewhat yellowish. It also has small black spots, six claws, and two legs on each segment from there to its end. Here, it rests on the uppermost gooseberry. It pushes off its skin several times and becomes entirely light green, without spots, like the caterpillar crawling from the leaf onto the lowest gooseberry; the pushed-off skin is lying on the leaf above it. After one or two days, it turns back to the color it was before. At the end of May it goes into its change and makes a yellow cocoon like the one lying on the leaf at the top, above the caterpillars. It remains there for two, sometimes three weeks, until the middle of June. Then a yellow fly comes out of it, of the kind seen flying downward near the top, which also likes to stay near the *Kloster-* or gooseberries.

Chapter 25: *Macaria wauaria* (V-moth), *Nematus ribesii*
(gooseberry sawfly) and *Ribes uva-crispa* (European gooseberry)

The V-moth appears on a fruiting gooseberry plant, which Merian notes is
eaten by numerous types of caterpillars. Her account of the caterpillar's defensive behavior of "standing upright" and thrashing violently as if angry is coupled with her image demonstrating the firm grip the insect has on the leaves of
the plants. She then describes its more typical locomotor movements, which
are characteristic of the Geometridae and are also discussed in the commentary for Chapter 12. Here she includes another behavioral observation, writing
that the nocturnal V-moths fly swiftly.

The other 'caterpillar' included on the upper gooseberry is not a lepidopteran larva, although it is easy to see how it could be mistaken for one. This is the
gooseberry sawfly, whose larvae decimate these plants just as she says; the species (*ribesii*) is named for its food (*Ribes*), as is the case for the naming of many
insects. Merian describes the voracious behavior, appearance and pupation of
this larval Hymenopteran, and she mentions the color of the adult fly, which
she observes staying near the gooseberries. The latter point is yet another field
observation that she includes, adding to the ecological content of her book.

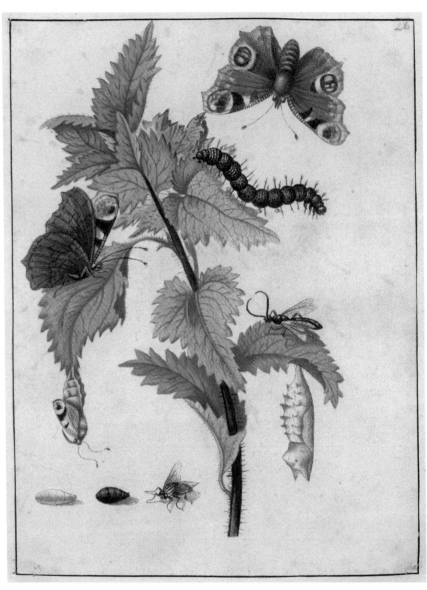

PLATE 26   Merian, 1679. *Raupen*

XXVI. Great stinging nettle *Urtica urens, major*

(53) No one would have convinced me at the time when I got a caterpillar like the black one on these upper leaves that such a beautiful summer-bird would come from such an unshapely form. But I have since learned that often something quite lovely has come from many an unattractive caterpillar. This type of all-black caterpillar is found in June, in no small numbers, on great stinging nettles, eating them so bare that nothing is left save the stems; hence, I could hardly give them enough nettles, so ravenous and active were they in feeding. I placed them in a large box, and one time when I forgot to clean it out I noticed on the bottom, underneath their droppings, white maggots or worms, some of which had already changed into black capsules. All of this took place within two hours, during which the maggots had nothing to eat. From this I observed that such things originated in their droppings. A light-colored maggot, along with the dark capsule, can be seen side by side at the bottom. Within twelve days, a common fly emerged, seen resting next to them. But after the caterpillars have eaten enough, they enter into their transformation and hang from their hind end, firmly attached to the inside of the box cover or to some other wooden surface, with head down, and thus change in the space of (54) four or five hours into a date seed that looks rather like a strange head or face. Its color was reddish, but the large date seed hanging from the green leaf is entirely yellowish; the other, half-emerged one is reddish. From all of this, one can see how a summer-bird is formed when it emerges. For it has quite small wings at first and is lighter in color; but after half an hour its wings are stiff, in their full size and natural color, so that they can fly away with them. The time in which it turned from the date seed into such a lovely insect was sixteen or seventeen days. I used to call it simply the peacock butterfly, since it has four such colorful eyes on the tops of its four wings. Its body, like the underside of each wing, is hair-colored, except that the wings also have many black stripes and spots. It has brown eyes, bright and transparent as glass, two brown horns, and four yellow legs. The rest of the upper wing surfaces is a pretty dark red. But on one occasion, and to my mind in a disorderly manner, there came from the above-mentioned yellow date seed a black, ugly, flying insect which had a foul smell. One such is resting on a green leaf. Once again, where this disorderliness comes from, I leave to the naturalists alone to judge.

Chapter 26: *Aglais io* (peacock butterfly) and *Urtica dioica*
(common or stinging nettle)

Merian found the larvae of the striking peacock butterfly[1] to be as ravenous as
they were unattractive, and was hard pressed to keep them supplied with nettles. She mentions no other food source for these caterpillars, and indeed, they
are specialists, feeding primarily on nettles.[2] In Plate 44, she shows another
species of butterfly larva that feeds almost exclusively on nettles, but she varies
the images by showing two different types of nettles that are abundant across
Europe, the common or stinging nettle and the small nettle. These distinctions
may seem trivial, but she accurately shows subtle differences in leaf shape of
these similar plants (the former has narrower leaves), and she correctly applies
the Latin names of that time to each plant species.

As in her preface and in Chapter 7, she records the appearance of flies whose
larvae fed on the lepidopteran frass, noting this happened when she forgot to
clean the box housing the caterpillars. She follows this with the observation
that when the caterpillars had "eaten enough" they changed into pupae within
four to five hours and attached themselves head down to a wooden surface.
Her use of wooden boxes "with a number of small holes" is mentioned again in
Chapter 30, where she describes this as her procedure for raising caterpillars of
all sizes and numbers.

Merian appears to have observed this butterfly's metamorphosis with particular care, and she recounts the emergence of the adult in some detail, picturing part of the process in her composition. Interestingly, she refers to it as
the peacock butterfly, the present common name of this insect, and one of the
few insects that she designates with any name at all. The bright colors and bold
eyespots on its wings made this butterfly a popular insect for inclusion in still
life paintings of the period. As she did for the comma butterfly, she depicts the
adult peacock butterfly in two positions to illustrate the very different pattern-
ing of the upper and lower surface of the wings, something that she also mentions in her text. Her use of lights and darks in the etched plates are especially
striking in this plate (Figure 74).

---

1   This species was previously called *Inachis io*. Plant and animal taxonomies are constantly
    being revised as evolutionary relationships are further examined.
2   Audusseau et al., Plant fertilization interacts with life history: 1–15. Another species that
    Merian features on nettles, *Aglais urticae*, is also a specialist feeder on this type of plant.

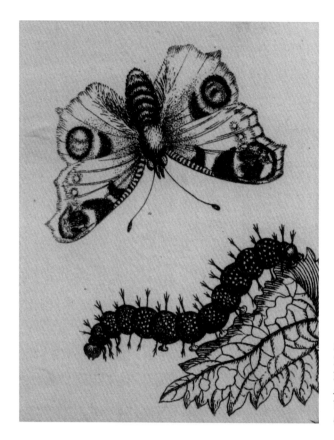

FIGURE 74
Detail from plate 26 of
Merian's 1679 *Raupen*.
Aglais io (peacock
butterfly) adult and
caterpillar

Here again in this chapter Merian records an instance of a 'disorderly' change, in which a foul-smelling and ugly fly came from the pupa, and she suggests that other naturalists should offer insight on this unexpected phenomenon. We now know this as some type of Ichneumonid wasp, several genera of which are common parasitoids of such butterflies. Her images cannot be used to identify these small flies or most of the other parasitoids to species; these insects comprise numerous species that can be challenging to classify even using a magnified photograph.

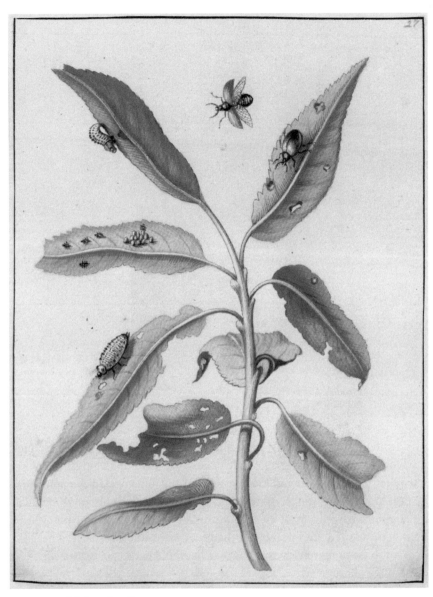

PLATE 27   Merian, 1679. *Raupen*

XXVII. Red willows   *Salix, acuto folio*

(55) At the same time that I was very carefully studying these little animals on the red willows that grow along the water, I would often find their eggs (on the leaves of the same) piled up together as if for a game of bowling. I also noticed that the little animals, stirred by the warmth of the sun, chewed a small hole in the top of the egg and crawled out like black, shapeless lice. Both they and the eggs can be seen on a straight leaf which has been broken or eaten off at the tip. All of this occurred at the beginning of June. They had six legs and immediately took those leaves as their food, but only the most tender parts, or the juicy green on the surface, so that nothing but the bare membrane was left behind. This they did for as long and as many days as needed to attain their full size, as illustrated in the little animal or worm crawling along the leaf bent downwards. These are yellow in color and marked all over with black spots, but the head is all black. Thus, on the fourteenth of June, they attached themselves to the middle of the leaves, hanging head down, and turned into date seeds, (56) almost unchanged in color, except that the head acquired more black, as can be clearly seen on the date seed hanging from the upper, darker-green leaf. Now after these seemingly lifeless date seeds had hung for fifteen or sixteen days without moving, at the beginning of June, little beetles crawled out of them which had an attractive dark ochre color and a pretty, black head like the date seed, also six black legs and two horns of the same color, as one will see in the beetles, one in flight and one crawling [down the leaf], at the top. This is one of the most thought-provoking and elegant transformations that will give anyone reading it cause to investigate further, or can elicit all manner of discourse from others hearing of it. I could perhaps add my poor opinion on it as well, but since that is not part of my undertaking, I shall leave it at that.

Chapter 27: *Chrysomela populi* (Pappelblattkäfer or red poplar leaf beetle) and *Salix caprea* (goat willow)

Merian's interest in plant-eating insects other than caterpillars reappears in this chapter. She describes the larvae of these small beetles as eating all but the "bare membrane" of willow leaves, leaving behind what a horticulturist today would call skeletonized leaves. Her statement that the lice-like larvae chewed their way out of sun-warmed eggs and immediately fed on "only the most tender parts" of willow leaves is indicative of her careful observations. This larval pattern of feeding is characteristic of many herbivorous insects, although the parts of the plants eaten may vary with the insect's developmental stage. We now know that in plants like willows, the nutrient content of the younger leaves is more favorable for the insects than that of the older leaves.[1] Her plate pictures the larvae, both tiny and large, a pupa, and extensive leaf damage to the young leaves near the end of a branch.

Her study for this image (study journal entry 7b) shows only the crawling adult and not the flying beetle, but it does include a narrow leaf on which the eggs are arranged like bowling pins, just as they appear in the final plate and in life. An error in describing the timing of metamorphosis occurs in the first edition version of this chapter; her text says that pupation begins in mid-June, but then states that the beetles emerge at the beginning of June when she meant July. The one-letter difference in the German spelling of these months may have led to a printing error, and this mistake was corrected in later editions.

Merian ends the entry on this "most thought-provoking" change with the prediction that others will discuss and investigate it, and the assertion that her "poor opinion" on such a transformation is not meant to be part of her book. One might suppose that she did indeed have ideas about transformations that were outside her experience with lepidopteran metamorphosis, but in the 1679 *Raupen* book, she consistently projects a modest persona. Throughout her European insect books, she shies away from speculation and adheres primarily to what she directly observed. In her 1705 *Metamorphosis*, her voice gains confidence, and the tone of her writing is more authoritative.

---

1   See for example Ikonen, Arsi. 2002. Preferences of six leaf beetle species among qualitatively different leaf age classes of three salicaceous host species. *Chemoecology*, 12(1): 23–8. The youngest larvae tend to feed between the small leaf veins, leaving behind the skeleton of the leaf, whereas older larvae and adults eat more of the leaf, tending to avoid only the largest veins, which are the most difficult to chew and digest.

PLATE 28   Merian, 1679. *Raupen*

XXVIII. Red-flamed rose   *Rosa versicolor*

(57) Now I turn once again—for the fourth time—to the rose, which the two kinds of caterpillars shown here like to eat; in particular, the lovely variegated roses on which I found both of them in June and with which I fed and maintained them through that entire month.

The first type of caterpillar is unshapely, small, and chestnut-brown in color. Behind its head is a bit of black, and it has six brown little claws, but otherwise no legs, rather it crawls on its belly like the snakes. Its form can be seen at the top, on the variegated rose. When it finally had its fill of food it entered its change, wrapped itself in a green leaf, and therein became a date seed that was brown in color and which I have placed on that same rose, to be more clearly visible. Now in such a transformed shape the caterpillars remained all of fourteen days, after the passage of which such pretty moth-birds crawled out, black and white in color. One can be seen at the top, resting opposite its caterpillar on the rose. (58) The second caterpillar is pictured at the bottom, crawling up the stem of the rose. It is green and has a brown head, a small stripe the same color directly behind its head; underneath, six legs of the same color and two more at its hind end, holding itself fast. I found it likewise in June on these variegated roses and maintained it on them. It went to rest for its change at the beginning of July and turned into a red date seed, having first pushed off its skin bit by bit and left it hanging behind, as both are seen on the lowest green leaf. Then, after fourteen days, a charming little moth-bird, speckled somewhat the color of wood, came forth and is seen flying downward at the bottom. One time, however, I observed one of these caterpillars that lay down as if dead, and soon thereafter a white maggot crawled out of it, such as lies on one of the green leaves wrapped around the stem. It changed directly into a brown capsule, from which the fly resting near it emerged after twelve days. All of this is sufficient to show how all kinds of roses have great enemies in the caterpillars. If these last-described caterpillars were the only ones, it would already be enough for them; but daily experience with the great damage they cause teaches every hard-working gardener that there are many more kinds than these.

Chapter 28: *Notocelia* sp. (Tortricid moth), *Pyraloidea* moth,
Tachinid fly and *Rosa* sp. (rose)

In the fourth *Raupen* book plate with a rose as host plant, Merian again
shows two species of caterpillars, another Tortricid leaf roller near the top of
the image and down lower, a moth from a different family, the Pyraloidea.[1]
Regarding the latter, she includes the description of a white maggot emerging
from a seemingly dead caterpillar, and the plate depicts what is probably a
parasitic Tachinid fly with its larvae and pupa.

She concludes the chapter with further emphasis on the wide variety of
caterpillars that are enormously damaging to roses. The four roses in this volume are of different types, strengthening her point that none of these valued
garden flowers is spared the ravages of caterpillars. However, Merian modified her rose compositions in various ways to avoid repetition and to maintain
the interest of gardeners and others who might wish to purchase the book. In
this plate, Merian included a small amount of leaf damage when compared to
the extensive damage to the rose shown in Plate 19. In all four rose plates the
leaves, buds, and flowers are shown from various angles and in slightly different stages.

---

1   This very diverse superfamily of insects, sometimes called "snout moths" includes leaf rollers
    as well as groups with very different habits.

PLATE 29   Merian, 1679. *Raupen*

XXIX. White currant *Grossularia hortensis*

(59) In early June three years ago, in a distinguished garden here, I noticed that this type of caterpillar had done much damage to the small red and white currant bushes planted along both sides of the walks, to the extent that almost nothing but the caterpillars themselves was left on them, since they had also chewed off the stems of the unripe berries, which lay in great number rotting on the ground. I have also found these caterpillars on ordinary hedges of black-thorn and hawthorn, but very few of them. They display a remarkable way of moving, because of their few legs: In front they do have the six black claws, but after that, instead of the usual eight middle legs, only the last two, round and yellowish; and at the hind end, their usual two, which (like the head) are black. So when they want to move about on them, they bring the two hind legs up to their front claws, and the rest of the body arches up. They are not easy to exterminate, for they lay their eggs well concealed in the remaining green leaves where they are not noticed. (60) The eggs are yellow and some-what smaller than millet, as can be seen here. But the caterpillars go into their change in the middle of July, taking the form of an oblong cocoon. Afterwards, at the end of July, their not unattractive summer-birds emerge. One thing can-not be passed over in silence here: namely, that it very rarely happens that the date seed resembles the caterpillar, or the caterpillar the summer-bird, in its color. Rather, the opposite is the case, so that from a black caterpillar will come a white summer-bird, and from a green will come a brown, and so forth, as will also be shown in the transformations to follow. This does not ignore the fact that the caterpillar described here and shown crawling down the middle is white, yellow, and black; the date seed's covering is also yellow and black; the summer-bird yellow, white, and black; as also the eggs are yellow. However, those with an understanding [of such things] know that nothing is detracted from the general rule by one example, and that such an unusual occurrence is to be considered exceptional and not typical.

Chapter 29: *Abraxas grossulariata* (common magpie moth) and *Ribes rubrum* (white currant)

In a rare mention of the place where she collected specific insects, Merian details the devastation caused by the magpie moth caterpillar on white currants in a "distinguished garden."[1] Perhaps she meant to underscore that fact that even the 'best' gardens were plagued by insect damage. She notes that almost nothing was left of the plants, and that even the unripe berries had been excised by the caterpillars and were rotting on the garden walk. Despite this, she made the aesthetic decision to show an uninjured plant rather than the skeletal remains described in her text. To inform gardeners, she reports that the eggs of this moth are well-hidden, making extermination difficult. As it is impractical to attempt to show cryptic eggs in her image, she places the small mound of eggs to one side, detached from the plant.

This is the third Geometrid caterpillar in the book, and she again details its curious form of locomotion. In this plate she emphasizes the absence of the "middle legs" (prolegs) and the use of the front and hind "legs" for locomotion by showing the caterpillar bridging the gap from one cluster of white currants to another further down the plant.

Although Merian did not attempt to classify insects, she did take note of patterns of appearance in her study animals, and in particular the coloration of insects in all stages of their life cycle. Here she remarks that most often, the various stages of lepidopterans have very different coloration from each other, but that here is an exception. All stages of the magpie moth cycle contain yellow coloring, and three stages have further similarities with black and/or white added to the yellow. This chapter provides a good example of the necessity of reading Merian's text along with her images, because both are needed to understand the depth and breadth of her extensive observations.

---

1 White currants are a variety of the same species as red currants. The garden mentioned could have been that of the Imhoff or Volckamer families, with whom Merian had connections through her pupils and other means, but there were many such gardens in Nuremberg at the time. She did not mention this garden in her study journal notes for this chapter (study journal entry 49).

PLATE 30   Merian, 1679. *Raupen*

XXX. Sallow, or pussy willow   *Salix caprea, latifolia*

(61) There now follows in our sequence an attractive type of caterpillar with something that sets it apart from the others: Namely, while other kinds move under the green leaves of trees so as to be better shaded from the sun and not be even completely dried out by it, enjoying rather the juice and moisture on the underside of the leaves; this one, by contrast, places itself firmly on the outer surface of the leaves and likes the sunshine rather than shade—perhaps because they are of moister nature and therefore demand more drying. For this reason, they prefer to stay high in the trees, and can easily be seen from some distance away, especially because of their lovely white color. For when it is illuminated even more by the sun, we notice them right away, and they appear to shine forth among the green leaves on account of their light yellow color. Their eggs, which are bluish green, I have shown above the crawling caterpillar, that is, on the stem of a green sallow or pussy willow leaf, which is where I found the eggs in May and on whose leaves the caterpillars love to feed. (62) These, however, are hairy on both sides, lighter-colored and more blunt than other willow leaves. Now from the eggs, little by little, these same beautiful caterpillars came, with black head and six black claws, eight yellow legs in the middle of the body, and then two more of the same on the hind end. They are wood-yellow in color and display very light yellow or white spots which (it is said) can be seen from afar. But I kept all my caterpillars, large or small, many or few, in a box, having first made a number of small holes in it so that they could have air. In the same container, on the seventh and eighth of June, they changed into date seeds that were black in color and had on top round, white spots with several small, white hairs on them. One is shown resting on the bent-down leaf. After they had remained in their change for two weeks, entirely white moth-birds came forth with four black-and-white speckled legs and a shimmering luster that looked like mother of pearl. One is shown at the top, flying downward. Moreover, like the silkworms, they mate right away and lay their bluish-green eggs as described above. After that, they live a few days longer and then gradually die, almost before one notices.

Chapter 30: *Leucoma salicis* (satin moth) and *Salix discolor*
(pussy willow)

This plate shows part of a damaged willow tree as the host of the satin moth, which like many insects, takes its modern scientific name from the plant on which it feeds. Merian noticed behavioral patterns in her subjects and comments that these caterpillars are unusual in preferring the outer surface of the leaves and warm sunshine rather than the cooler, moister underside of leaves where she found most types of larva. The text adds a good bit of information that is not in the handwritten study journal entry for this species, e.g., the entire section on moisture preferences and her description of the egg location. Nor does Merian mention the wooden boxes in her study journal entry. Differences between the two texts give us insight into what she considered important information for her readers.

The satin moth text includes a rather complicated description of the colors of the adult moths, referring to them alternately as white and bright yellow. In this species the adults range from pure white to white tinged with yellow gold, in part depending upon the light in which they appear, and this may be what she was attempting to accentuate in her text. Alternatively, it could be that she is referring to the colors of the larvae, but this seems less likely, as caterpillar color is described in the second paragraph. However, she then mentions the adult color as being like "mother-of-pearl" adding to the confusion about her meaning in the first paragraph. Interestingly, her study journal shows two adults, one a male with the narrower abdomen and more feathered antennae, and the other a female with a larger abdomen, figured above the eggs and in profile (Figure 75). Her choice in the final composition was to show the wider-bodied female in flight and to omit the profile view.

In this plate, she pictured the eggs near the top of the plant, a placement similar to that in Plates 20, 21, 27 (a beetle) 30, 37, 45, and 47. Whether this was an aesthetic decision or meant to represent where she most frequently observed eggs is not known. However, it is the case that in nature, many species of lepidopteran deposit eggs where they will be exposed to sun, and naturally, this location tends to be near the top of plants or at the ends of stems and branches. Merian ends the entry with yet another connection to the behavior of related species by noting that these moths, like silkworms, mate "right away," lay their eggs, and die in a few days after emerging from pupation.

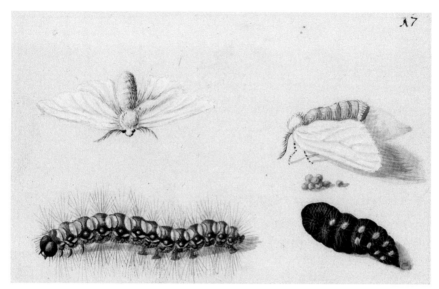

FIGURE 75   Merian. Study painting 17 of *Leucoma salicis* (satin moth)

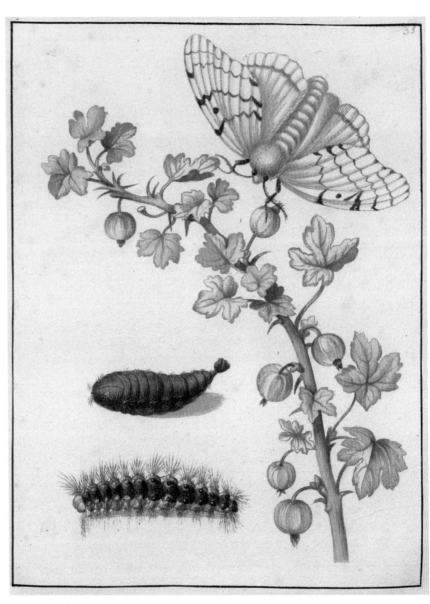

PLATE 31   Merian, 1679. *Raupen*

XXXI. Small, white gooseberries

Grossularia alba, vulgaris

(63) One might suppose, as I myself once thought, that a beautiful caterpillar would make a handsome moth- or summer-bird; and conversely, that an ugly caterpillar would make an ugly insect. But I found, little by little and with wider experience, that my assumption was wrong; indeed, that most of the time the opposite occurred. Because many a time, from what I considered an ugly caterpillar a very pretty insect grew, and from a pretty caterpillar a very ugly insect. This was clearly seen in the preceding chapter, where the caterpillar was a lovely white and yellow and gleamed brightly when the sun shone on it, but from which only a small, plain moth-bird came; so that if its whiteness had perhaps not been compared to gleaming mother-of-pearl, it would be considered merely a poor and unattractive moth-bird. Thus it is with the caterpillar under consideration here, which on first inspection is quite pretty, with a wide stripe like velvet down the length of its back, and on that stripe, beginning at the head, twelve blue grains, that is, six on either side, like (64) beads of turquoise; and from there on, almost to its hind end, six more grains on each side, but which are red; and at the very end, two more blue grains, as above. From all of them little gray hairs protrude. But the rest of the body is ash-colored, the head and legs entirely yellow. I found this caterpillar at the beginning of July on the small gooseberries, which it took for its food and one branch of which is shown here. They are also found on rose, plum, sour cherry, and all kinds of trees, to which they did great damage this year. Before the end of August this type changed into brown date seeds with little yellow hairs on the outside, having first pushed off their skins, which, shrunken very small, remained attached to the end of the date seed. Both can be seen just above the caterpillar. If such a date seed was touched, it would twist about so endlessly that one had to marvel. In September there finally emerged from it a large but none-too-beautiful moth with a fat head and black eyes, as can be seen on the one in downward flight. The horns and legs were also black. The rest of the moth is white with brown and black flecks and stripes. They fly only at night, like owls [*Eulen*], that cannot see by day; for that reason the Dutch call them simply *Eulgen*.

Chapter 31: *Lymantria dispar* (male gypsy moth) and *Ribes uva-crispa* (European gooseberries)

This second appearance of white gooseberries shows what may have been a wild plant with smaller fruit, as opposed to the cultivar with much larger fruit pictured in Plate 25. Merian's investigative focus was on her insect subjects, but in picturing their host plants she presents a sampling of what was growing in Nuremberg and environs in her time. Many of the cultivated trees and flowering plants were brought from abroad, and this may have been the case for what she terms a "Spanish gooseberry" in the earlier plate.

She discussed the life habits of this moth in Chapter 18 and so takes this opportunity to muse on the relative beauty of caterpillars and the appearance of their adult form. She was intrigued by the trend that she has observed, in which there is frequently an inverse relationship that cuts both ways. In the instance of the gypsy moth, what she deems a pretty caterpillar gives rise to an ugly adult. Her accurate and detailed description of the common gypsy moth caterpillar, another foe of gardeners, concludes that with its velvety stripe and beads of red and turquoise, this devastating pest is lovely in its own right. She repeats the information from Chapter 18 regarding their broad range of food choices and the severe damage that the insatiable caterpillars inflict on plants. Lastly, she likens the less than beautiful moth to an owl due to its night-flying habits, and introduces her readers to the Dutch word for moths, *Eulgen*. This is an interesting departure from her usual adherence to German names for the insects, and indicates that she was garnering information from a wide array of sources.

PLATE 32 Merian, 1679. *Raupen*

XXXII. Common meadow grass

Gramen pratense, vulgare

(65) The large caterpillar shown here at the bottom is one that I have regularly found on the grass in July, hence I assume that this is the best food for it and the one that it prefers. It is gray in color, and on the three segments directly behind the head it has black bands like velvet running downward; but when it contracts, there is nothing to be seen of them. Furthermore, it has brown spots on each segment all along its back, and on the last one, a rather short, brown horn. Also, on the lower part of each segment are warts or little knots above the legs. All of this is carefully pictured here. Now when the time of its change comes and it has found a suitable place, it pushes off the skin of its entire body and makes a gray cocoon, oval in shape, around itself, and inside it takes the form of a gray date seed. After that it remains resting alongside its discarded skin into August, when a large moth comes out which looks very much like the moth in Plate XVII, though that one was shown in flight, while the present one is shown resting, at the top. Otherwise, both moth-birds are of a dark orange (66) color and have very large heads with black horns and six legs of the same color. They fly only at night and lay white eggs with sea-green spots, which can also be seen behind the insect. The date seed with the discarded skin lies directly below it. The cocoon lies above the caterpillar and looks as if it were spun of finely cut wool, but thinner or somewhat more transparent from the outside. At this point the attentive reader or observer may note (just incidentally) that the colors of all moths or moth-birds (be they of one color, or however they might appear) are not durable or, we might say, permanent colors, but merely like the purest fine-cut wool which is then artfully strewn on them, as if Nature had dusted it on or (so to speak) applied it in fine detail, to give such an insect a far lovelier appearance. For if one were simply to brush off the wings, there would be nothing left but a very thin creature of isinglass and not at all pretty. Of summer-birds it will be noted that their color is also easily brushed off, but it more resembles very finely ground flour than the wool just described. This I have taken into account, and I think it could be a noteworthy distinction. Now if someone wishes to ponder all of this further and will apply his thoughts a bit to the way that God, through his handmaiden Nature, adorns many an inconspicuous and (as we might think) useless thing so wonderfully and beautifully, he will on every hand find occasion to practice his devotional meditations the better for it.

Chapter 32: *Gastropacha quercifolia* female (lappet moth) on grass

This entry on the female lappet moth is linked to one on the male lappet featured in Chapter 17, where she notes the similarities of the adults. The caterpillar in Plate 32 is the third different color morph that she depicts for this species, and as in Chapter 17, she takes special care to describe its appearance. In Chapter 32, Merian focuses on the female lappet and shows the moth's eggs and describes their color. She again explains how the exoskeleton or 'skin' is pushed off the last-instar caterpillar's body as it turns into a pupa, and she illustrates this structure clearly in the plate.

Of interest in Chapter 32 is her discussion of the basis for the colors of lepidopteran wings, made more intriguing by providing further evidence that she used a magnifying glass in her observations. She compares the substance of the moths' wing colors, "like the most finely cut wool," to that of butterflies, which she remarks is more like "finely ground flour." As she says, neither type of coloring is permanent or embedded into the wing, but can be brushed off, leaving the thin and colorless infrastructure of the wing. Merian likens the appearance of such denuded wings to isinglass, an odd but apt comparison. Isinglass is a clear gelatin made from sturgeon and other fish, and it can be used for glue.

Her study journal entry on the lappet mentions the appearance of the moth's coloring, but it contains no comparison with that of the butterflies, something that she found important enough to comment on when she compiled the text for her book. By designating this a "noteworthy distinction" she draws attention to her discovery of this difference in coloration. In her next sentence, she deflects notice of her astute observations by suggesting that others ponder these details and, through such study, better appreciate how God's handmaiden (Nature) has made such beauty. Decades later, in *Metamorphosis*, Merian refers to the scales of butterfly wings as looking like roof tiles,[1] another apt visual analogy. We might assume that by the time she wrote that entry (around 1702), she had access to an improved system of magnification, perhaps a more powerful hand lens or a simple microscope.

---

1 Merian, *Metamorphosis*, text with Plate 9. "Seen through the magnifying glass, this blue butterfly has the appearance of blue tiles, looking like roof tiles lying in a very orderly and regular manner."

PLATE 33   Merian, 1679. *Raupen*

XXXIII. Blackthorn   *Acacia Germanica*

(67) In this caterpillar crawling across two blackthorn leaves I discovered an unusual characteristic: While others feed only on leaves that may contain moisture and therefore have no need of water, these are very inclined to drink, as I have often refreshed them with water. From this I judge that their food is mainly the cause, being of a dry and astringent kind such as the leaves of the blackthorn, the ones on which I have found them singly each time. This year, however, in which the beginning of spring brought a long period of heavy, though not cold, rains followed by warm and then mild weather, I have found many thousands of such little caterpillars on quince and plum trees. Now when this large number, compared with the individual ones from before, caused me to wonder, I took great pains to investigate where they had come from. And thus I discovered that many hundreds of eggs had been laid encircling several of the aforementioned quince and plum tree branches (as can be seen on the stem in the illustration), some of them just skins, for the caterpillars had already crawled out; from which I conclude that either the previous, exceptionally hot summer must have lent them greater strength or that the abovementioned uncommon, temperate rainfall, the likes of which we have not had in the five-year span of my investigations, could have helped considerably in their maturation. (68) Now as soon as they emerged from their eggs, they gathered on their particular branches in the evening and constructed a large, thick cocoon where they remained into the month of July. They were then transformed, having also attained their full size in the meantime. But before that, each of them made for itself another, denser yellow cocoon, like the silkworms, in which they turned into a black date seed. One can be seen lying at the bottom, but the cocoon hangs from one of the twigs. Now such a cocoon must be able to withstand cold well, considering that there has not been so cold a winter in a long time; the snow cover remained over a quarter of the year without any melting. The color of these caterpillars is white on top, then yellow on both sides, bordered by two narrow lines and, in the center, one wider black line, followed by blue and finally yellow again. Their six small claws were black, the remaining legs yellow. Otherwise, they are very slow and lethargic. From their droppings grew (as happened with others, too) white maggots, of which one is seen crawling on the round plum. They had soon spun themselves into little brown capsules, one of which is lying just above it on the twig. After fourteen days common flies or gnats came forth. But the moth-bird that was produced is of a light liver color and is very active at night. The newly laid eggs are yellow at first, but later brown, just as also happens with other eggs, particularly those of the silkworms.

Chapter 33: *Malacosoma neustria* (lackey moth) and *Prunus spinosa* (blackthorn or sloe)

Merian was captivated by the lackey moth caterpillar's habit of drinking water. She compared this thirsty caterpillar to the majority of larvae that she observed, which obtain moisture only from their food. She wrote that the sour plum leaves the caterpillars preferred were "dry and astringent," a statement that echoes Goedaert's writing on the same species of moth. However, Merian considers drinking by caterpillars to be unusual, because "others feed only on leaves that may contain moisture and therefore have no need to drink," whereas Goedaert thought drinking a more common behavior. He noted the egg-laying pattern of the lackey moth, wherein the female moths deposit their eggs in rings around a twig, but he did not include an image.[1] Merian's plate prominently featured a ring of lackey moth eggs circling the branch of a well-chewed blackthorn (or sloe) tree, and an outstretched caterpillar nearby, greedily consuming leaves from the same tree. It appears that she initially etched the eggs on a leaf of the plant, but later obscured these to a great extent (Figure 76), and she does not mention them in her text.

In this chapter, Merian again comments on the enormous population of insects, presumably referring to the summer of 1679, which she mentions in Chapters 10 and 18 as being temperate and favorable for the caterpillars. She noted that she saw "many thousands" of these small caterpillars, and in actively searching for their source, she found "many hundreds" of eggs (some already hatched) encircling the stems and branches of quince and plum trees. As in Chapter 18, she attributes this abundance to favorable weather conditions that she describes as unusual in the five-year course of her study, remarking that in a more typical year she found the caterpillars singly rather than in great numbers. Merian appears to have been one of the earliest to consider the effects of weather on the numbers of insects appearing in a given year. She addresses the 1679 population boom again in Chapter 41, and she predicted that a large number of caterpillars would appear in 1680 because of all the eggs being laid.[2]

---

1 Goedaert, *Metamorphosis Naturalis*, Volume 1: 36. He writes that the caterpillars eat dry and astringent food (a type of willow) and that they drink water. In his second volume, Goedaert states that many caterpillars drink. Ibid., Volume 2: 55.

2 Réaumur has often been credited with being the first to publish on the effects of climate on insects, but clearly he had predecessors in this; moreover, he cites Merian in his work, so he was familiar with her books.

FIGURE 76   Detail from plate 33 of Merian, 1679 *Raupen* book. The eggs encircling the stem
are characteristic of *Malacosoma neustria* (lackey moth). Eggs on the leaf above
the caterpillar may have been added in error and then partially covered by the
addition of more etched lines.

As described but not pictured in the plate, the caterpillars of this species form
a group web where they congregate, leading to the term 'tent' caterpillars.
Merian's account of the timing of pupation is not clear. She implies that they
overwinter as pupae, but this species typically overwinters in the egg stage. As
in earlier entries, she documented the appearance of a fly whose larvae grew
on the caterpillar droppings; she refers to these as "common flies or gnats."

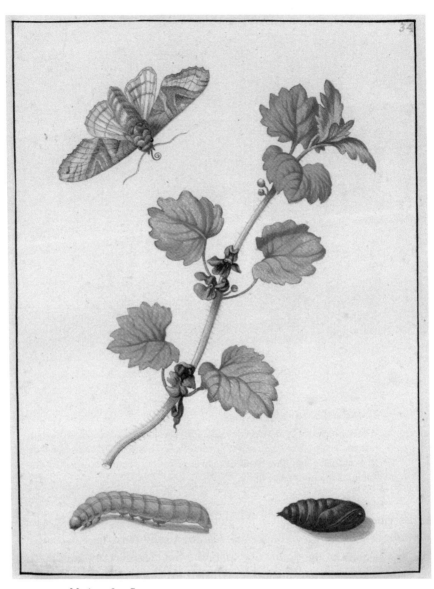

PLATE 34   Merian, 1679. *Raupen*

XXXIV. Ground ivy in flower

Hedera terrestris, florens

(69) Although one might think that I had deviated somewhat from my proper sequence in the preceding transformation by including the fruit of the wild plum, it should be understood that I did this intentionally. Considering that blossoms were no longer present at the time of that caterpillar's first transformation, I chose to include a branch of unripe, green plums rather than one with just leaves. Also, in years past I have had the caterpillar much later than this year, so the bluish-green fruit of the plum is appropriate. But all the other caterpillars in this little book follow one another according to when, from month to month, they were accustomed to spinning themselves up. The same is true, in almost every case, of the flowers and other plants available for them to feed on at such times, with the exception of the first page showing the mulberry tree, which is placed at the beginning of the book for the reasons noted there. This next type of caterpillar, which appears here in the correct sequence, I have always found in July taking as its usual food the above-named ground ivy with its blue blossoms. They are by nature (70) very timid, for whenever they sense or feel the slightest thing, they immediately roll themselves up and remain still until certain they have nothing to fear. Then they return to their feeding. It has happened that one or the other caterpillar (one is seen crawling at the bottom of the page), when it did not have its proper food, would also feed on wild plum leaves, but as soon as they could get their usual food again they would pay no more attention to those. They entered their first change, or transformation, at the end of July, turning into dark brown date seeds, one of which can be seen next to the caterpillar. At the end of August such lovely moth-birds came out as the one shown in downward flight here. Their body, along with the two hind wings, is of a wood-yellow or light yellow-ochre color, but the two forewings are green and somewhat reddish, also pale yellow. The six legs are the same wood-yellow color. They fly only at night, thus they are not often seen or found.

Chapter 34: *Phlogophora meticulosa* (angle shade moth) and
Glechoma hederacea (ground ivy)

Not until chapter 34 does Merian elucidate the rationale for the sequence of
the chapters in her book, which was the rough chronological order in which
the insects pupated throughout the spring and summer. For this reason, the
book begins with spring plants in flower and proceeds to later-flowering plants
and plants in the fruiting stage. Most of the plants shown by Merian in this
book are either cultivars or common meadow plants and would have been well
known by her readers, especially the gardeners among them. In this chapter,
she justifies her choice of the fruiting stage of the wild plum for her previous
plate, which shows it with the lackey moth. She then goes on to explain that
the angle shade caterpillar is found in July, thus clarifying the times at which
her insect subjects appear in nature.

The larval defensive behavior of the species is described, and she character-
izes these caterpillars as very timid, rolling up and remaining motionless until
a threat has passed, whereupon they immediately begin to eat again. This caterpillar is typically a generalist feeder, but Merian attributes to them a preference for what she terms ground ivy. She pronounced the adults difficult to find
because of their nocturnal habits, but then again they are cryptically colored
and not easily seen when still. By showing the adult in flight, the details of the
moth's wing pattern was clearly displayed in the etching (Figure 77).

FIGURE 77   Merian. Detail from plate 34 of Merian's 1679 *Raupen*. *Phlogophora meticulosa*
(angle shade moth)

PLATE 35   Merian, 1679. *Raupen*

xxxv. Blue lilies, or garden iris   *Iris hortensis, latifolia*

(71) I have often found many of this curious sort of caterpillar, one of which is pictured here, in August, on the broad, hard leaves of the blue lilies on which they fed. For although these caterpillars are quite soft, they can eat up those same hard leaves very quickly, and it has often amazed me more than a little that they should have such strong chewing parts in their mouth. Now if it happens that the caterpillars do not have this kind of green leaves, they will devour each other, which I discovered one time when I had forgotten to give them their accustomed food. I found a number of them that had earlier spun themselves up (yet still looked like caterpillars) and had been eaten away. But as soon as I offered them their usual food, they abandoned them and ate their leaves as before, as shown by the crawling caterpillar: They begin at one side of the green leaf and eat all the way across to the other side; then they begin again at the top, and so forth. Now this type of caterpillar is green in color and marked with darker green all over, very quick when moving about. I had them (72) for three years in a row, yet only with great effort did I get a few date seeds, but no insects from them, since (and I do not know what caused it) they were either dried up or otherwise spoiled and thus could produce no insects. Until finally in the fourth year of my investigations, in February of last year, to my great delight, this single one, namely a moth, came forth. It was brown and had a white spot on each wing. The rest of it was very nicely and distinctly marked with dark brown. This type of moth-bird flies only toward evening or at night. There is also this to note: The caterpillars described here, unlike some others, very often pushed their skin down over themselves (the same thing occurred with them at the time of their transformation) then left it hanging on their date seed, as one can see on the chestnut-brown one resting on the curled leaf. The insect in downward flight is also clearly pictured there.

Chapter 35: *Melanchra persicariae* (dot moth) and *Iris* sp. (iris)

The caterpillar of the dot moth amazed Merian with its ability to feed successfully on the stiff leaves of an iris (which she also calls a blue lily). She credits the
larvae with "strong chewing parts" and emphasizes the distinctive pattern of
herbivory employed by these caterpillars in both word and image. Her description and the lively composition combine to give her readers a clear picture of
exactly how these insects feed and an understanding of their voracious nature
(Figure 78). The dot moth caterpillar is another species she notes as cannibalis-
tic when deprived of plant food. The caterpillars are described as very quick in
their movements in contrast to those of the moth in the next chapter.

The difficulty of obtaining adults from caterpillars is part of the story in this
chapter. Her persistence is highlighted by the fact that it took her four years to
obtain the entire life cycle of the dot moth. She was successful in getting a few
caterpillars to pupate during the first three years, but no adults emerged from
what she termed "dried up" or spoiled pupae. In the winter of her fourth year
working with this species, an adult dot moth emerged. Merian's delight in this
success with such a small and dully colored moth seems almost equal to her
pleasure in obtaining the much larger emperor moth with its "beautiful shadings and contrasting colors."[1]

---

1   Merian, 1679. *Raupen*, Chapter 23.

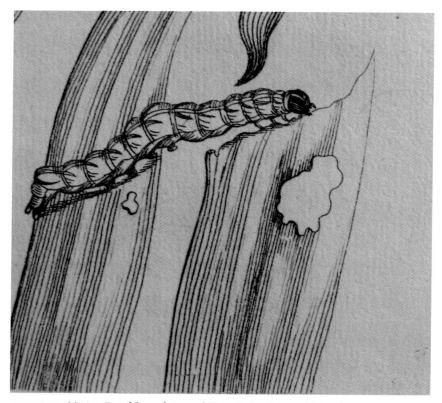

FIGURE 78   Merian. Detail from plate 35 of Merian's 1679 *Raupen. Melanchra persicariae*
(dot moth) caterpillar feeding on iris leaf

PLATE 36   Merian, 1679. *Raupen*

XXXVI. Broadleaf plantain   *Plantago major*

(73) Having often found these pleasant and attractive green caterpillars on the broadleaf plantain, and noticed that the same little herb was the food they preferred to eat and fed heavily on it, I was very pleased that such a beneficial, glossy little plant, which many people crush underfoot in the grass or on the street, should find a place here in what is to be my little book, which many a painter, because of such a beautiful drawing by another hand, could perhaps use in the foreground of his own landscape painting or similar work. Thus I fed this caterpillar (which is very slow in its movements) on such small plants into the month of August. At times they also got by with various kinds of lark-spur, which I gave them when no broadleaf plantain was to be had. But as soon as they tasted the plantain again, they lost interest in the other flowers and their foliage. Now when these caterpillars were finally satisfied and had attained their full size (one is shown crawling on the flowering plantain), they very quietly went to rest for their first transformation, gradually shriveled up, and turned into completely brown date seeds. The image of one rests on the lower leaf. (74) After they had remained lying immobile inside it for a time, up to December, insects came forth which were entirely dark brown, but otherwise not very attractive, though on their two forewings they are marked with a white circle, and they have six brown legs and two black-brown horns, as the moth resting at the bottom shows. Concerning this moth-bird, I should repeat just a few words from my commentary on Plate XXXI concerning the many ugly moth- or summer-birds that can develop from beautiful caterpillars. Thus I refer [the reader] back to that chapter, since the caterpillars I have just described appear pleasant and attractive to our eye, while the moth-birds that result from them lack nearly all graceful charm.

Chapter 36: *Lacanobia oleracea* (bright-line brown-eye) and
Plantago major (greater plantain)

As with the dandelion composition earlier in the volume, Plate 36 features
a humble and overlooked plant, "which many people crush underfoot." The
caterpillars of the bright-line brown-eye moth eat plantain, but they are generalists that feed on a wide variety of herbaceous plants. Merian gave them
larkspur when she could not get plantain, but noted that they preferred the
latter and would go back to feeding on plantain whenever it was offered.

She wished to depict the entire above-ground portion of this particular
plant, as she did with the dandelion in Plate 8. In both compositions, she reduced the scale of the plants to show the entire above-ground growth form
within the size of the quarto plate, while maintaining the insects as life-size.
It is unlikely that a plantain large enough to flower and set seed would be so
small; Merian used artistic license to convey added information to her readers about lesser-known plants. For the more familiar cultivars figured in other
plates, the parts of the plants she depicts are usually shown life-size, requiring
that their images be cropped to fit the plates for this book. As in Chapter 8, she
left us a clue to her motivation for featuring this plant, referring to "a beautiful illustration by another hand" that could inspire other painters to use the
plantain in their own landscape foregrounds. She may have been referenc-
ing the iconic "Large piece of turf" painted in 1503 by Albrecht Dürer that
features both plantain and a dandelion intermingled with meadow grasses.
Dürer's small painting is considered a seminal work of Renaissance naturalism, but the plantain appears in larger paintings by him as well.[1] The plantain could have been seen by Merian in other Dürer paintings, or in works
by his various emulators such as Hans Hoffman (see for example Figure 34
on p. 55).

The slow-moving caterpillar, considered beautiful by Merian in contrast to
the rather plain moth, became inactive before pupation, an accurate observation, and one that she also mentions in Chapter 2 regarding the clay moth
caterpillar. In nature, the bright-line brown eye would overwinter as a pupa,
and the adult stage would emerge in the spring, but she reports emergence in
December. In the next chapter, she addresses the effect of temperature on the
unnatural winter emergence of adults, indicating that she realized this was a
result of keeping the insects in her relatively warm house.

---

1 Koreny, Fritz. 1988. *Albrecht Dürer and the Animal and Plant Studies of the Renaissance.*
   Boston: Little, Brown and Company.

PLATE 37   Merian, 1679. *Raupen*

XXXVII. Large red sour currant   *Grossularia hortensis, majore fructu rubro*

(75) In my discussion of this transformation I must repeat at the outset what I stated in the twenty-ninth chapter regarding the type of caterpillar described there: that is, how rarely it happens that a caterpillar, its date seed, adult insect, and eggs are all the same color. Indeed, I must confess that I have almost never since observed such a similarity as there, four times the same color, on account of which that complete process of transformation can rightly be considered an exceptional rarity. Here follows now another beautiful caterpillar quite similar in color to its attractive insect, but its date seed and eggs are of a different color. Thus, what that scholar wrote: that as the caterpillar is, so the rest will be, is in fact rarely the case. Something other than a similarity of its color must have been meant, but because he included mention of the color, I cannot well leave it unchallenged. Now I found this lovely caterpillar in August on the large red sour currants, whose green leaves I then provided for its maintenance. It was bright white in the sunlight, but brownish in the shade, and had, on each segment down its back, small black spots in a diamond pattern; and it had a very strange (76) way of moving, because (although there were three of the usual claws on either side of the first three segments behind its head, which is split on top and appears like two heads) it had only two round legs on the third-from-last segment and two more at its very hind end. Now when it wanted to move, it would hold fast with the hindmost part of its body and raise itself up high until it could grasp something with its front claws. Then it would bring the hind part of its body up to the claws and advance in that way, which one could hardly watch without delight. The form of this caterpillar can be seen attached to the large green leaf by one hind leg and leaning with its front claws on the stem of the leaf. But it changed, or (properly speaking) was transformed into a reddish-brown date seed at the beginning of December. It is also shown lying on the large leaf, and it remained unchanged until February of the following year, when a beautiful moth-bird emerged that was (as I have said) not unlike its caterpillar: entirely white and marked with black spots. After several days it expelled its eggs, which were light green; all of which can be seen on the aforementioned green leaf. But I believe that because I kept this caterpillar's date seed in my living quarters over the winter, the warmth noticeably helped the earlier birth force[1] of the insect; since otherwise, if it had lain still in the cold in some open field or other place, perhaps it would have come out later.

---

1   This is literal; Merian does call it "birth" (*Geburts-kraft*).

Chapter 37: *Biston betularia* (peppered moth) and *Ribes rubrum* (red currant)

Continuing to note patterns of coloration in life stages of insects, Merian compares the peppered moth to another Geometrid species, the magpie moth of Chapter 29. Both have caterpillars that can be the same color as the adult moths, although in actuality, caterpillar pigmentation can vary a great deal based on a number of factors, as mentioned previously. In fact, caterpillars of the peppered moth are particularly adept at matching the surfaces on which they live, and remarkably, can reverse coloration if moved from a dark surface to light, and then back to dark.[2] Unusually for Merian, she contradicts the writings of an earlier unnamed scholar, possibly Jonston,[3] saying that she cannot leave unchallenged his observation that all stages of a moth or butterfly tend to be colored like the caterpillar. She notes that this is rarely the case and that the peppered moth is an exception to the usual pattern of differently colored stages.

Here again, she notes the distinctive locomotion of the caterpillars of the Geometrid family and adds that the looping motion could hardly be watched "without delight." In her composition, she repeats the motif of showing the larvae bridging plant structures with their bodies. Merian described their characteristic notched head shape, supplementing through text something that is difficult to see in the life-size image of the small insect. The adult female is shown laying eggs on the top of the plant (Figure 79). She also describes oviposition by the female in her text, although eggs are not mentioned in her study journal. The eggs were painted yellow in the counterproof plate shown in this volume, but she described them as pale green.

After mentioning the onset of pupation in December,[4] she discusses the fact that the adult emerged in February, attributing this to the warmth of her house and reasoning that out in a cold field the appearance of the adult moth would have happened later. Merian's writings regarding the effects of temperature on

---

2 The precise triggers of the exceptional phenotypic plasticity of caterpillars like those of *Biston betularia* are still being investigated. The peppered moth caterpillar is capable of mimicking colors of its environmental background regardless of food source, and becomes dark when raised in black plastic containers or light in white containers even when fed the same food in both enclosures. Noor, Mohamed A. F., Robin S. Parnell, and Bruce S. Grant. 2008. A reversible color polyphenism in American peppered moth (*Biston betularia cognataria*) caterpillars. *PLOS ONE*, 3(9): 1–5.

3 Jonston, 1653. *Historiae Naturalis de Insectis.*

4 This would be abnormally late for pupation to begin and may have been due to her captive animals experiencing unnatural photoperiods and temperatures.

FIGURE 79   Merian. Detail from plate 37 in Merian's 1679 *Raupen. Biston betularia* (peppered
moth) laying eggs

the rate of change in insects was insightful, because she was comparing what
she observed in nature with what happened under artificial conditions. These
final comments on the effect of temperature on metamorphosis are absent
from her study journal, providing another example of a general conclusion
that she forms after years of observations.

PLATE 38   Merian, 1679. *Raupen*

XXXVIII. A type of garden fennel

Fœniculum hortense

(77) This lovely and nicely banded caterpillar is one that I came upon in August, on the fennel which it took as its food (pictured here with the caterpillar on it). Such caterpillars are a pretty green in color and have black bands, like velvet, and golden yellow flecks on the bands. If they are touched hard, they immediately extend two yellow horns on the front of their head, like a snail. They also have in front, that is, underneath, three pointed legs or claws on each side, then two more segments without legs, and then four more segments with four round legs under them on both sides, then two more empty segments, and at the very end, two more round legs with which they hold on very tightly. In case they have no fennel, they will also eat yellow turnips. Gardeners call this caterpillar *Oebser* [fruit-eater],[1] since they believe that it does great damage to fruit, although I have never found it on anything but (as I said) fennel and yellow turnips. It does have a peculiar odor though, like fruit when many different (78) kinds are mixed together. Now when they have attained their full size, they push off their entire husk, or skin, which remains attached to them, as I have pictured at the top; and hang from a wall, head down, attaching their hind end as securely as if it were glued there. In the middle of their body they spin a white thread around them so as to remain firmly suspended. Then they change within half a day into date seeds like the one I have shown here hanging from the fennel, having the form of a swaddled infant on which one could almost recognize a human face. These date seeds are gray in color, and partly also green. They hang in this form into April or May, although some have also crawled out for me in December, but I attribute that to the fact that I kept them in a warm room. The summer-bird shown having come forth here has four wings, and the two upper wings are a beautiful yellow and black, the two lower ones also, save that they are a beautiful blue in the spotted areas or fields, including the lowermost oval, though it also shows some red. The body is also black and yellow, it has six black legs and, at the front of its head, a long beak which it rolls up completely, and if sugar is put out for it, it will place the extended beak on the sugar as if to eat with it. It also uses it to draw the sweetness from flowers, as I have observed many times.

---

1 The name is presumably derived from German *Obst* (fruit).

Chapter 38: *Papilio machaon* (swallowtail) and *Foeniculum vulgare* (sweet fennel)

Swallowtail caterpillars begin to appear in late summer, and here Merian depicts an adult perched on a lacy fennel plant, which she designates as its preferred food. Swallowtail larvae do eat fennel, but will avidly consume several other plants. In yet another departure from the painted study in her journal, the finished plate includes the everted 'horns' that appear when the swallowtail caterpillar is threatened. The peculiar odor she mentions is emitted from these scent glands. What we now know are defensive structures, the 'horns' were described in her study journal, where she wrote that the caterpillars will extend them if they are touched with some force. As in several other entries, the "legs or claws" of the larvae are described in some detail, and she adds that the swallowtail caterpillars can attach very tightly to a surface with the hind "legs."

Her journal image also differs in that the adult is shown in flight, whereas in the published composition she presents a side view of the perched adult. For every other butterfly in the 1679 *Raupen* book, Merian presented the adults in two positions, most likely in order to show differences in wing coloration on the two sides. Unlike some other butterflies, the swallowtail has very similar coloring on both sides of its wings, and she may have posed it this way to emphasize its tongue, or 'beak' as she calls it. She understood its use for feeding on plant nectar and explains that this proboscis can be extended or rolled up. Goedaert writes about the use of a "tongue" in two other species but not in the swallowtail butterfly.

A third difference from her journal study appears in representation of the pupae. Oddly, Merian's plate shows the pupa head down, and she writes about it hanging in this position. This head-down position of a pupa would be atypical for the species, and she did picture it in the head-up position in her study journal. Perhaps the pupa had no plant material to which to attach in her wooden boxes. In this case she would not have observed how the swallowtail pupae are attached to plants by a silken thread, angled away from the stem. Some butterflies do pupate in the head-down position, and she may have been swayed by this fact to illustrate the swallowtail pupa this way. As with many late summer butterflies, the swallowtail overwinters as a pupa, and this was reported by Merian. Once again, she notes that some adults emerged in December and attributes this to the warmth of her house.

PLATE 39   Merian, 1679. *Raupen*

XXXIX. A kind of garden mint   *Mentha hortensis, verticillata*

(79) This next caterpillar in the sequence, green and white speckled, I have found on the leaves of the beneficial and fragrant garden mint, a little herb generally known here as *Herzentrost*, one type of our familiar and generally named *Deimenten*, as true mint or, in other places, balsam is called. Here in this town, large quantities of the latter wild but fragrant plant are spread before the doorway in honor of a newly married couple on the first day of their wedding celebration. With the leaves of the former plant, however, I have maintained these caterpillars into July. When one touched them, they would thrash rapidly back and forth with their head, as if they were very angry, and they would do this a good ten swings at a time. Then after a while they would feed again and quickly, as if they were in a great hurry. One of these caterpillars is shown resting on two of the green leaves of the first mint named above, which is still young and without blossom, since the caterpillars have eaten it. After that, in mid-July, they went into their change and made (80) a thin cocoon around themselves, in which they were transformed into dark brown date seeds, the image of which I have placed lower down, on a green leaf. They then remained in this form and appearance, lying quite still, into the following month of August, when moths came out (one of them is resting on the ground) with heads the color of oranges. But their bodies, along with the two horns and four legs, were brown. On top, or where the light shone brightest, they appeared as if gilded; their wings, too, all shone like the most beautiful gold. But if one turned them onto one side, they appeared greenish. On each wing there were also three brown spots, and the wings were brown as well underneath, toward their tips. These moth-birds, as I have often pointed out, fly only at night, but by day they are quite still. Like other flying insects that circle around a burning lamp until they finally burn themselves up and, with their mutilated wings and singed bodies, become food for the spiders and other vermin.

Chapter 39: *Diachrysia chrysitis* (burnished brass moth) and *Mentha* sp. (mint)

Plate 39 differs from her typical compositions in showing the plant growing from the ground. Merian includes information on the local names of the plant and recounts the Nuremberg tradition of using the fragrant mint to decorate the doorway of newlyweds. She indicates that the caterpillars ate both the leaves and blossoms of the mint, and did so as if "in a great hurry," but added that if disturbed, the caterpillars would thrash about and swing their head rapidly as though they were angry.

The cocoon and pupa look like a cutaway view as do some others in the book, but it is more likely that she was attempting to represent the pupae inside what she described as a "thin cocoon." The burnished brass moth pupa provides a good example of this, and it is the case that in this species, the cocoon is so thin as to be transparent, revealing the dark brown pupa inside.

She noted the usual information on their metamorphosis, and the adults' behavior of night flight and quiescence during the day. Interestingly, her description of the adult, confusing on first read, contains an early description of what we now know as iridescence in the wing color of the moth. The common name of this moth comes from its appearance, which is gold "on top, or where the light shone brightest" and green when seen from the side. This manifestation of 'thin film interference,' a term for a type of phenomenon that causes iridescence, is known in numerous insects as well as other animals. Robert Hooke and Isaac Newton were pursuing the structural basis of this interaction of light and surfaces around the same time that Merian made her observations on the changing colors of the burnished brass moth.[1] She ends the entry by describing a simple example of an arthropod food chain – flame-singed moths provide food for "spiders and other vermin."

---

1 For a description of the phenomenon in a variety of animals and the history of study see Ball, Philip. 2012. Nature's color tricks. *Scientific American*, 306(5): 74–9. For more details on the structural basis for the phenomenon in lepidopterans, see Ghiradella, Helen. 1991. Light and color on the wing: structural colors in butterflies and moths. *Applied Optics*, 30(24): 3492–500.

PLATE 40   Merian, 1679. *Raupen*

XL. Blue garden larkspur   *Consolida Regalis,*
hortensis

(81) The beautiful moth-bird shown here is one I have often observed on blue
larkspur blossoms. Because it was obviously so lovely and of such unusual
color, I often wondered what kind of caterpillar it might originate with or
come from. Then, after diligent investigation, I was able to find its caterpillars on those very same flowers, to which they did great damage: for they not
only favor them as food but often eat blossoms and leaves so completely that
nothing remains but bare stems. In color they are attractive and beautifully
speckled, and it is a marvel and a delight to see them: white and black and with
a yellow stripe along the lower body. One hardly notices them when they are
resting on larkspur, since the leaves of the plant are so long and curled; one
must search carefully for them. One of the caterpillars is shown here on a blossom between the green leaves at the top. These caterpillars are also somewhat
slow-moving. At the end of July they (82) went into their change, or proper
transformation, and in a very short time became date seeds of an attractive
brown color (one of which is resting on a green leaf at the bottom). In that
form they remained lying there until May, sometimes June, of the following
year. Only then did such lovely moth-birds emerge, which were also somewhat
lethargic or quiet in their behavior, adorned with a beautiful rose or purple
color with white bands. They have six legs of the same color. One of these
moth-birds is represented here in the middle of the copperplate resting on two
green leaves—the more so to delight the eye of appreciative admirers and to
shed light on one small masterpiece of tireless Nature.

Chapter 40: *Periphanes delphinii* (pease blossom moth) and
Delphinium sp. (larkspur)

Merian expressed delight with the marvelous colors of both the larva and adult
of the pease blossom moth, and something like admiration of the caterpillars'
ability to eat the larkspur down to the bare stems. The species, like many lepidopterans, is named for its preferred food plant. The rendering of the moth's
legs is more detailed in the finished plate than in the study journal painting
(Figure 80).

   This entry emphasizes her persistence in finding and collecting the larvae
for this striking moth and alludes to the caterpillars' ability to blend in with the
curled leaves of the host plant. She remarks on the slow movements of both
larvae and adults, and she correctly categorizes the pease blossom moth as
an insect that overwinters in the pupal stage. The entry ends with the suggestion that her composition was designed "to delight the eye" and to highlight a
diminutive but masterful work of "tireless Nature." As discussed above, this reference to 'Nature's art' may have been inspired by any number of writings on
natural theology, but also perhaps by an engraving made by Matthaeus Merian
(Figure 27 on p. 44).

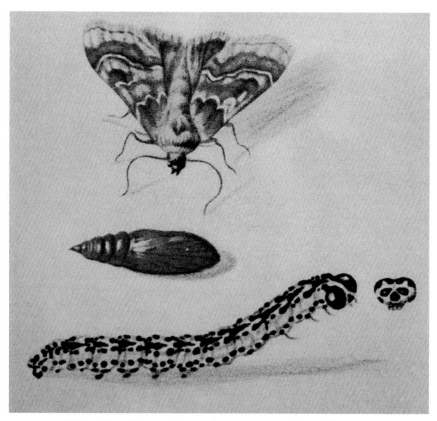

FIGURE 80 Merian. Detail from study journal painting 38. *Periphanes delphinii* (pease blossom moth)

PLATE 41   Merian, 1679. *Raupen*

XLI. Lesser wild orach   *Atriplex sylvestris*

(83) The little caterpillar shown here raising itself up on the stem of a green leaf near the center is somewhat greenish in color, like a leaf that has nearly dried out. I have found such ones on the lesser wild orach in August, when it blossoms, and kept it fed on the plant up to the end of that month. It has a most curious way of moving, holding itself by its hindmost legs and searching (by rearing up with its body and front legs) for where it can go next. Unlike the other caterpillars, it has no legs under its midsection. Now at the end of the aforementioned month of August, it changed into a brown date seed, the form of which I have shown lying on the lowest drooping green leaf. So it remained until April of the following year. At that time a small moth-bird came out which was of an even uglier green color than its caterpillar. One is shown here flying toward the blossoming plant. At this point it will not be inappropriate, given the brevity of this chapter, for me to reply to and discuss a question recently put to me, (84) namely: what I think of the many thousands of uncommonly large caterpillars we have had, and in some measure still have (particularly in this 1679th year's generation); and if this might not be a sign of future ills. To which I, in my female simplicity, offer the modest reply: It is already plain to see, by the nearly bare fruit trees and other damaged plants, that this voracious plague of caterpillars, just by itself, bodes no good for us. But one can easily imagine that the eggs of the already present summer-birds, or of those yet to emerge, will be found again in great numbers on those same trees and in other places, too. Yet time will tell, especially the spring of next year, when damaging cold or an extended period of rain at the time of their changes and growth could easily destroy them. But thanks be to God that for the present, even with such great numbers of caterpillars, all that was lost is being restored with rich blessing, and we ourselves are made truly glad thereby.

Chapter 41: *Eupithecia* sp. (pug moth) and *Atriplex* sp. (orach or saltbush)

In sharp contrast to the previous plate, Merian presents here a drab moth that she termed even uglier than the caterpillar, and she pictures its life stages on its rather common and monochromatic larval food source. This Geometrid moth is probably the plain pug moth, *Eupithecia simpliciata*, whose larvae feed upon *Atriplex*, as well as *Chenopodium*, a very similar plant in the same family. It is not possible to be sure of either the insect or the plant species from Merian's image; a magnifying glass is usually needed to identify these insects precisely, even from living specimens.

She may have written this chapter before she recorded the descriptions of the other four Geometrid caterpillars. Merian recounts its "most curious way of moving" as if describing this for the first time, and remarks that "unlike the other caterpillars" it has no feet (prolegs) under its midsection. The pug moth may have been the first caterpillar from this family for which she made careful observations of locomotion, but by virtue of the insect's appearance in August, it became the last Geometrid caterpillar featured in her book. Let us recall that Merian raised insects over a period of at least five years before assembling the text and images into a book, and the order in which she made her careful observations is impossible to determine. She may have edited the text describing each insect after determining their order of appearance in her book, but that appears not to be the case with her account of the specialized style of caterpillar locomotion exhibited by the Geometrids.

Whatever the order in which Merian composed the entries on this distinctive group of larvae, she did note each time she discussed them that they were unusual in lacking the middle 'legs' of most caterpillars. Yet she talked about each of the Geometrid caterpillars individually and did not group them together in any sort of classification scheme. Her lack of interest in ordering her study subjects into groups makes her something of an outlier at a time when classification was beginning to gain importance in natural history. She instead concentrated her attention on how the insects lived, moved, and fed, and used the entries in her book to create pictures of their lives. In this chapter, she conveys some of her particular interests by focusing on the extraordinary movements of the larvae. With words and text she paints a scene that showcases a tiny upright caterpillar that most people would never notice. Here in Plate 41 the action of the scene is centered on the larval insect, clinging to its perch with its rearmost prolegs and searching in space for the next leaf or stem by which it will pull itself up.

Because the caterpillar description was brief, she digressed in this chapter and once again addressed the issue of the 1679 caterpillar population boom, and what it might foretell. This passage, another broad conclusion that is absent from her study journal, gives us further insight into the way Merian wishes to portray herself to her readers. On one hand, she presents herself as expert enough to have been asked a question about the phenomenon, but on the other, she demurs that she can offer only a "modest reply." Her reference to her "female simplicity" seems somewhat disingenuous, because she clearly thought a good bit about climate effects. In this chapter, she speculates that the next generation of insects could be in great numbers if the weather were to favor survival of the various life stages. She ends her discussion by thanking God for providing rich blessings in spite of the fruit trees stripped bare by hordes of caterpillars.

PLATE 42   Merian, 1679. *Raupen*

XLII. The black poplar tree *Populus nigra*

(85) This is a twig from the black poplar tree, also known as the *Alberbaum*. When the leaves are still young or small they are a lovely green color, and if one touches them they will stick to one's fingers. There is also a yellow salve made from them, which is used on the hair. Now very often there are round lumps which grow directly on the leaves or on the stems; and if one cuts them open before they are mature,[1] they are light green and entirely empty, as I have observed with particular care through the magnifying glass. However, when they are ripe, they take on a beautiful bright red color, burst open, and in all of the lumps—in each one, as in the others—are found six kinds of living creatures, as clearly illustrated in an opened one on the lower part of the stem. The first two types of these creatures are very small flies or gnats [*Mücklein*], one type smaller than the other, and they bite like gnats [*Schnacken*]. But the other four types are crawling creatures, each one smaller than the next; indeed, one is as small as a dot or point. (86) I have placed all of them next to the round water droplet, seen gleaming brightly in the form of a small, round pearl in the middle of this burst-open lump. This water droplet does not dissipate, but remains this way, whole and round unless one disperses it forcibly. But the largest crawling creature, or worm, which can be seen in the above-mentioned lump and again on one of the leaves at the top, takes all of the others as its food. For it lies very still, and when the other creatures crawl around it, it grasps them with its beak and sucks out all their moisture, so that nothing is left but their skins. Then it takes another, until it finds no more. Thereupon it moves onto a leaf and is changed into a date seed like a light-colored bubble in appearance (such as can be seen on the just-mentioned leaf, directly below the worm I was describing—and three small flying creatures of the first type as well) and remains for seven to twelve days lying or hanging that way. Then there emerges from it an insect, in form like a wild bee, its body yellow and black and its head dark red, with six yellow legs. One is shown resting on top of a leaf opposite the one just described. Whether all these creatures grow out of the tree's moisture, or how they came to be inside, these are questions I leave to those who are better versed in Nature.

---

1 Merian's word here is *zeitig*, a term she also applies to a completed stage in insect metamorphosis.

Chapter 42: *Populus* sp. (poplar) with gall insects

Plate 42 is unusual in several ways, the first of which is Merian's focus on the plant material depicted. Naturalists as early as Theophrastus and Aristotle were intrigued by gall formation and by the creatures inhabiting galls.[2] Her fascination with galls is apparent in the eighth study journal entry, in which she used a piece of vellum three or four times the size of her usual study image to create a detailed painting of a poplar branch with large swellings on its petioles (Figure 81).[3] She described in both the study journal and the published text how she made her observations, with the additional mention in the latter of the use of a magnifying glass. In the study journal, she mentions studying galls again in 1684 and in 1710, and explains that she dissected them in the company of a distinguished physician.[4] The information she provides on her examination of the galls over a period of years gives us further insights on her field and 'laboratory' studies, some of the techniques she used, and her unflagging curiosity about nature.

In this chapter, she again defers to others on the question of how the insects come to be inside the gall of the poplar tree. It would in fact be centuries before the questions as to how galls form were answered. We now know that they are abnormal plant growths triggered in reaction to invaders such as the ones that Merian depicts in Plate 42. She states that there were six types of creatures in the mature gall, although she did not see anything in the early stage, even with her magnifying glass. The insects pictured in Plate 42 are difficult to identify with certainty, but they appear to be a collection of gall midges (family Cecidomyiidae), gall aphids (possibly *Pemphigus spirothecae* and *Pemphigus populinigrae*), and a Syrphid fly that has predatory larvae. Merian wrote that the larvae of the fly sucked the other 'worms' dry, which is an apt description of the Syrphid's habits. Gall midges are so small and delicate that it is difficult to imagine how she preserved them in an intact state in order to paint them.

Her observation of the "small round pearl" inside the gall sounds mysterious until one understands that these tiny spheres are made by the gall aphids as a means of storing their excreted honeydew. *Pemphigus* aphids coat what would

---

2   Thanos, Costas A. 1994. Aristotle and Theophrastus on plant-animal interactions. In *Plant-Animal Interactions in Mediterranean-type Ecosystems.* New York: Springer: 7.

3   Galls may form on leaves and their petioles, on the stems of the plants and on buds. Merian's plate appears to depict galls formed by aphids (petiole galls). Gall midges cause similar pathologies, and she may have observed both types of gall insect in the poplar tree. In these cases, abnormal plant growth can be triggered by insects feeding on the plant or laying their eggs within its tissues.

4   Appendix, study journal entry 212.

FIGURE 81
Merian. Study journal
painting 8 of poplar
branch and insects

be a sticky mess with water-repelling wax, making the wax globules easy to
move, thereby saving space in the tight confines of the gall.[5] Gall aphids are
social, and the soldiers, who remove these globules from the gall daily, have a
different structure from the reproductive females; this may explain some of
the different types of creatures Merian observed. The waxy nature of the coating on these waste products was not described until almost 200 years after
Merian's book was published.[6]

5   It was not until 1876 that naturalists understood that these "marbles" consisted of wax. Pike,
    Nathan, Denis Richard, William Foster, and L. Mahadevan. 2002. How aphids lose their
    marbles. *Proceedings of the Royal Society of London B: Biological Sciences*, 269(1497): 1212.
6   Buckton, George B. 1883. *Monograph of the British Aphides*. London: Ray Society, Volume 4.

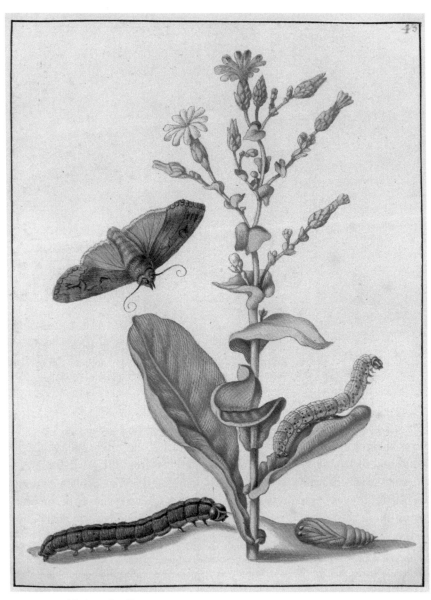

PLATE 43   Merian, 1679. *Raupen*

XLIII. Head lettuce in flower $\quad$ *Lactuca capitata, florens*

(87) This changeable type of caterpillar is one I have come upon by diligent searching in August, a form that was light gray below, with a wood-yellow stripe running down its entire back. One of them can be seen here with its body and hind legs resting on the rolled-up green leaf, but rising up with its front end and remaining legs. But not long after that, it changed its initial form, or shed its skin, and turned dark brown, which caused me some wonder. Its head, however, remained yellowish, like that of the larger caterpillar crawling along the ground. Its food was lettuce, an example of which is this flowering plant with its yellow blossoms. Others of this type also eat bindweed, which is a plant not at all favored by gardeners because of its habit of attaching itself everywhere, and if its roots are not checked in time, it will severely damage the entire garden with its rapid growth. I have also found these caterpillars in the ground, from which (88) they emerge only at night to search for food. At the end of August, the kind I am describing changed to a date seed, which was a pleasant light brown to look at. It is pictured lying next to the caterpillar on the ground. This date seed remained completely motionless into September, when the moth shown here in flight crawled out of it. Its head and two fore-wings looked brown, but the body and two hind wings were a light blue-violet. These moth-birds will not likely be seen flying at midday, but rather toward evening or at night, when they do appear and can be easily recognized by their rustling sound, like that of bats; and for this reason they are called bats in some places. On the other hand, insects that fly in daytime and do not have such large heads may well be called summer-birds, considering that, like true birds, they prefer to fly around by day and stay in their accustomed places at night; which, as I have said, these moths tend not to do.

Chapter 43: *Agrotis ipsilon* (dark sword-grass or black cutworm)
and *Lactuca sativa* (lettuce)

Two color morphs of the black cutworm are shown by Merian, who puzzled over the color change that occurred after the molt of their exoskeleton. This phenomenon was something that she appears to have been among the earliest to observe, and which she noted again in Chapter 50. As in other compositions, she shows these caterpillars 'in action' as it were, one lifting its upper body from a leaf and the larger one crawling on the ground. Her image supports her text, in which she wrote that the larvae feed at night on or below ground. Once again, this is an accurate observation for this species, and her careful field observations are borne out. Merian draws attention to her efforts by beginning the chapter with remarks on her 'diligent' search for caterpillars. The insects in the finished plate have just a few differences from those in the study journal painting (Figure 82); the coiled antennae may have been an artifact of preservation.

As with other caterpillars appearing in August, this species overwinters as pupae and the adult moths emerge in the spring. She pictures the caterpillars on lettuce, but perhaps to indicate that caterpillars could be the gardeners' friend as well as foe, Merian added that bindweed, a bane of gardeners, was also eaten by the cutworm larvae. The differences between moths such as the dark sword-grass that fluttered about in the evening or at night and day-flying summer birds is mentioned again, and in this entry, she revisits another distinction – that moths have larger heads than butterflies.

FIGURE 82   Merian. Detail from study journal painting 56. *Agrotis ipsilon* (dark sword-grass or black cutworm)

PLATE 44   Merian, 1679. *Raupen*

XLIV. Flowering / small stinging nettle

Urtica urens, minor

(89) When this caterpillar appears in August, it spins itself up in September and remains lying that way through the winter into March of the following year. At that time its insect emerges and is seen flying in our region during the spring. But when it happens that the caterpillar appears in May or June, it spends only fourteen days in its transformation and then comes forth as the same kind of insect. Its caterpillar is yellow with narrow and wide black stripes all over. It has six black claws, eight yellow legs in the middle and two more legs at the end. It pushes its skin off several times, and when it has attained its full size it crawls onto a wall and once more pushes its slough, or skin, completely off. After that, it hangs itself head down, attaches its body as securely as if it were glued there, and becomes a date seed. These are light in color at first, later a somewhat more brown wood color. Some may also appear as if gilded. If one tears them away, nothing comes out, and they wither. Their form can be seen, along with the aforementioned skin, hanging down from the green leaf. But in many a date seed very small, white worms sometimes grow—sixty or seventy of them—and afterwards the same number of small, golden flies crawl out of their capsules, two of which can be seen, together with two little worms, next to the date seed. But from a good date seed comes a summer-bird like the resting or the flying one—both are of the same kind. The upper side (90) of their wings from the outer edge is a bright red lead color; the rest is a lovely dark vermilion. They have both large and small black spots in which a beautiful blue can be seen. But the underside is dark yellow above, then the color of brown hair, nicely adorned with large and small flecks. The body and the four legs, along with the two horns, are brown. Now when one has many of the caterpillars together, a number of white worms grow in their droppings. One is shown lying on a small leaf in the middle of the nettle. And within the same day, these turn into red-brown capsules. Then, after fourteen days, a common fly comes forth, like the one sitting at the bottom, which has a red head, a dark yellow body, and six black legs. When the sun is fully up, they begin to fly like other summer-birds, though only singly; but after the sun has passed midday, more of them can be seen. With approaching high temperatures or humid conditions and thunderstorms, one will see them flying around in large numbers, lower and fluttering; then they can be more easily caught. In some places,

these summer-birds are called *Zwifalter*, where it is thought that they have only two wings, considering that the pair underneath is not easily noticed unless one has the insect in hand. Also, they are called *Goldvögelein* because many are colored such a beautiful, bright yellow, like gold; or *Buttervögelein* because some tend to that color. And when the sun's warmth lasts long, almost nothing but white and yellow summer-birds swarm on the meadows toward evening.

Chapter 44: *Aglais urticae* (small tortoiseshell) and *Urtica urens* (small nettle)

The description of the small tortoiseshell is longer than the average *Raupen* book chapter and contains a number of detailed observations on the biology of this butterfly. The length and depth of her account is of special interest, because this insect was the subject of the second entry in Merian's study journal, which is positioned under an inscription describing it as her first investigation. The abundant and naturally occurring tortoiseshell butterfly, whose larvae feed on nettles, would have been readily available to the young Merian. Additionally, tortoiseshell adults fly during the day and are colorful and obvious to the eye.

Most of the insects she studied are univoltine, reproducing only once per season, but her observations led her to understand that the small tortoiseshell was an exception to this pattern. She begins Chapter 44 with the information that this insect can overwinter as a pupa, but that it may also transform in the summer. The tortoiseshell is indeed bivoltine, meaning that it is capable of producing two generations of caterpillars in one year. One generation completes a full cycle in early spring and summer, and the second generation overwinters as pupa and emerges in the spring as an adult. The observation that *Aglais urticae* is capable of this relatively rapid growth and reproduction may have been original with Merian. In her 1679 book she included two other butterflies whose larvae feed heavily on nettles, *Aglais io* (Chapter 26) and *Polygonia c-album* (Chapter 14), but in Germany these two species are more typically univoltine, having only one generation per summer. It is therefore notable that she reported a second summer generation only in *Aglais urticae*, which is consistent with contemporary research.[1] The fact that there are two generations per season of tortoiseshell caterpillars foraging on an extremely common plant would have made it an easy species for the budding naturalist to collect and raise in her first empirical study.

Merian goes on to describe the shedding of the exoskeleton of the several instars, pupation of the tortoiseshell, and her attempt to observe what was inside the casing. This entry is one of eight in the 1679 *Raupen* to describe parasites/parasitoids of the larval or pupal stage of a lepidopteran. In this chapter, she pictures and describes what appears to be an Ichneumonid wasp of undetermined species, which she refers to as small, golden flies. Goedaert briefly

---

1 Audusseau et al., Plant fertilization interacts with life history: 3. Of the three nettle-eating species of butterfly, only *A. urticae* is bivoltine in southern Sweden.

described "black flies" emerging from one tortoiseshell pupa, while showing an uninfected pupa in the same plate (Figure 3 on p. 6).[2]

In addition to these parasitoids that kill the host, Merian mentions other flies, possibly some in the family Phoridae, which sometimes grew in the frass of the caterpillars (see also Chapter 7 and 26). The metamorphoses of these flies are also described, although in less depth than that of the butterfly. These accounts are included in her study journal, although there she did not paint the adult fly, but only its pupa. The study journal image also contains a great deal of detail including parasitic flies, and it may be that she added these particulars later in her life, rather than when she was 13 and just beginning her studies of insects.

Here as in four other compositions with butterflies (Chapters 14, 26, 45, and 48), she depicts both the inner and the outer sides of the wings. It is not until this late chapter in the book that Merian details various German terms used for butterflies. The extent to which Merian observed insects in nature comes through again in the later part of this chapter, as she recounts the timing of their flight and the effects of weather. She ends the entry with the poetic image of summer-birds in the warm evenings of long summer days fluttering about the meadows.

---

2   Goedaert, *Metamorphosis Naturalis*, Volume 1, Plate 77. Goedaert does not describe the colors of this butterfly, so it is not possible to be certain this is the same species as Merian's small tortoiseshell. However, it seems a close match in appearance, even though he states that the larvae eat elm leaves, which would be incorrect.

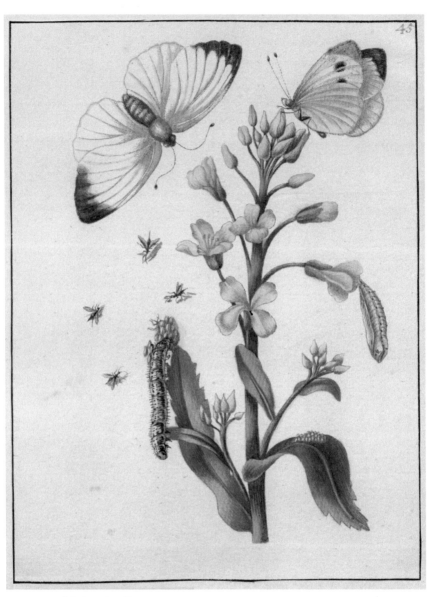

PLATE 45   Merian, 1679. *Raupen*

XLV. Yellow kale in blossom   *Flos brassicæ viridis*

(91) Almost everyone knows of and recognizes this type of caterpillar, which, because it is found in great numbers on all sorts of kale and cabbage, does much damage to them. Their color is black-and-yellow and green. They often eat the plants so bare that nothing remains of them but the ribs or stubs. They cannot tolerate cold, persistent rain but will be mangled by it and reduced to water so that nothing is left but their skins. Like those in my preceding chapter, they also have two times when they are transformed. If that occurs in August or September, they remain resting the entire winter until the following May and emerge as white summer-birds, one of which can be seen here in flight and another resting at the top. But if the caterpillars appear in June or July, they remain for twelve to fourteen days in their date seeds (one hangs from a blossom here) and then come out as the same white summer-birds. At times, one of these caterpillars will come to rest on a wall or even on one of the plants and remain quite still and without any food for several days. Then countless little white worms crawl out through its skin (92) and immediately spin them-selves into as many little yellow capsules. The worms and capsules both are pictured behind the caterpillar. Afterwards, the old one from which they came forth spins them all together, joins itself securely to them, remains another fourteen days without any food, and dies. But in twelve days from the time of their being encased, just as many little black flies come from the capsules, and these have lived only a few days for me. Now what the underlying cause of such different changes might be, and whether the mother caterpillar does not also model something in particular with her little ones, I leave to the judgment of other learned persons. Finally, it is noteworthy that such ones, like all the other summer-birds, not only alight on their food exclusively and fly mostly around such areas (which is why they are simply called cabbage birds in many places), but usually lay their yellow eggs there as well. So that, when their little caterpillars emerge from them, they will be well supplied with it and grow to greater size.

Chapter 45: *Pieris brassicae* (large cabbage white), *Pieris rapae* (small cabbage white) and Braconid wasps (Braconidae) on *Brassica oleracea* (wild cabbage)

Plate 45 appears to depict two different species of butterflies, both the large and small cabbage whites. In her study journal image, Merian shows only the perched small cabbage white. For some reason, she added the much larger flying adult to the plate, which by its size would appear to be the large cabbage white, a similar but distinct species. It is difficult to determine which species of caterpillar and pupa are represented, but the group of eggs would seem to be from the larger species, because the small cabbage white deposits fewer eggs in one place. This would seem to be the only entry in Merian's 1679 *Raupen* book that conflates two types of lepidopteran as one. The other oddity in this plate is the pupa dangling precariously from the end of a flower, something that would be unlikely in nature.

In other ways the chapter on these very common cabbage white butterflies bears similarities to the preceding entry. Both insects are bivoltine species capable of two generations in a season, which Merian describes as having "two times when they are transformed."[1] She details the timing of these transformations and, as in other species for which this is the case, remarks that they may overwinter in the 'resting' (pupal) stage. Like the tortoiseshell butterfly, both types of cabbage whites are heavily parasitized, and here Merian depicts and describes what appear to be Braconid wasps (possibly genus *Apanteles*) preying on the larval stage. *Pieris* larvae are common insect pests, decimating vegetable gardens of kale, cabbage, and related plants.[2] The widespread occurrence of these lepidopterans probably accounts for the fact that a number of other naturalists of the period observed the gruesome attacks of these parasitoids on cabbage whites.[3] Goedaert describes the parasitoids in much the same manner as Merian, but he did not illustrate them with the cabbage white. The

---

1 Goedaert also observed that this butterfly had two generations in one season. Goedaert, *Metamorphosis Naturalis*, Volume 1, Plate 11.

2 Many common cultivars are varieties of *Brassica oleracea*, including Brussels sprouts, broccoli, cabbage, cauliflower and more. Plants in the family Brassicacae are a leading food choice of many caterpillars as well as having been a staple in human diet for centuries. These herbivores were so numerous that it has been said that gardeners could hear the crunching sounds coming from the cabbage patch.

3 For a review of this history, see Van Lenteren, Joop C. and Charles H. Godfray. 2005. European science in the Enlightenment and the discovery of the insect parasitoid life cycle in the Netherlands and Great Britain. *Biological Control, 32*(1): 12–24.

similarity to Goedaert's wording in some of Merian's text both here and in her Chapter 22 indicates that Merian must have been influenced by his writings on the cabbage white butterfly. However, she would have had to observe the phenomenon herself in order to picture it in the detail shown in her study journal and in the etched plate.

The account of the Braconid wasps in this chapter is the most in-depth of any parasite description in the 1679 *Raupen* book. In addition to her report that that caterpillar ceases to move and feed, she adds text and image to show how the "capsules" of the little "worms" (Ichneumonid larvae) are spun up together, similar to her narrative in Chapter 22. In writing that after twelve days "just as many little black flies [Braconid wasps] come from the capsules," she appears to mean that one insect per capsule emerged. Again, she puzzles over this odd change and leaves it to the judgement of others. Almost 200 years after Merian's book was printed, Charles Darwin wrote to Asa Gray a strong statement as to how appalling he found this manner of reproduction. Darwin declared that he could not be persuaded "that a beneficent and omnipotent God" would have designed and created the Ichneumonidae "with the express intention of their feeding within the living bodies of caterpillars."[4]

Merian's final point in this entry is that the adult "cabbage birds" are named for the plants around which these butterflies congregate. She reasons correctly that the females deposit their eggs on the food source for their young in order to supply them with what they need "to grow to greater size." Goedaert mentioned this maternal behavior in his first insect volume. The adult insect he depicts is difficult to identify with certainty but may also be one of the cabbage white butterflies. He included the information that the caterpillars were damaged by rain, as does Merian.[5] In many such cases, her accounts went further than Goedaert's, and her images of the insects were better rendered. In Plate 45, she shows the eggs of the cabbage white on a leaf of the plant. Although these are the only butterfly eggs she included in her plates for the 1679 *Raupen* book,[6] she indicated that flying around the larval food source is a common behavior of the 'summer-birds.' The choice of egg deposition sites

---

4   Charles Darwin letter written to Asa Gray, 22 May 1860. Darwin, Charles. 1993. *The Correspondence of Charles Darwin*. Volume 8 (1860). Cambridge: Cambridge University Press: 223.

5   Goedaert, *Metamorphosis Naturalis*, Volume 1: 73.

6   Merian's moths probably laid eggs in captivity from time to time, because many moths mate soon after emergence. Butterflies on the other hand may mate and oviposit days or weeks after emergence, making it more difficult to match eggs with the adult species with any reliability. Merian shows eggs for 14 species of moth in the 1679 volume, but for only the cabbage white butterfly.

on larval food plants is now a well-studied component of plant–insect ecology, and Goedaert and Merian appear to have been the earliest to describe this critically important behavioral adaptation.[7] As she correctly deduces, by being born on their food source, the caterpillars will be nutritionally well supplied and will have a much greater chance of survival to metamorphosis.

---

7   For a review of this highly adaptive maternal behavior, see Renwick, J. A. A., and Chew, F. S. 1994. Oviposition behavior in Lepidoptera. *Annual Review of Entomology*, *39*(1): 377–400.

FIGURE 83   Merian. Study journal painting 83. *Pieris rapae* (small cabbage white)

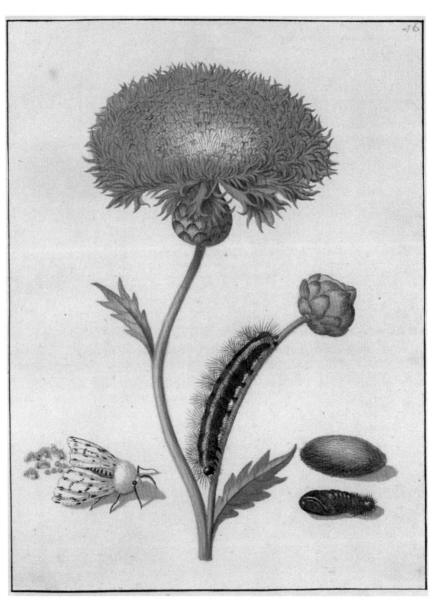

PLATE 46   Merian, 1679. *Raupen*

XLVI. Purple musk [bisam] flower

Jacéa moschata purpurea

(93) It is a source of wonder to me that I often had caterpillars which fed on a single flowering plant, would feed on that one alone, and soon died if I did not provide it for them. On the other hand, I had many other caterpillars that used more than one plant for their nourishment, though for a number of these one food was preferable. Indeed, they would move immediately from one food to another that they preferred as soon as they tasted it. Then again, I have had six or more types of caterpillars with a definite taste for an assortment of things—in fact, the same things—and which fed on them with equal pleasure, such that I was more than a little surprised by it. The present caterpillars also belong to this last group, inasmuch as I have found them in large numbers on all sorts of plants and flowers, peas, sugar peas, plantain, parsley, white nightshade, musk flowers (one of which is shown here), and many others. Such caterpillars had a broad, golden-yellow band along their entire back and were otherwise all brown, and covered with black hair. When they (94) are preparing to enter their change, they make from their hairs a brown-and-black cocoon in which they turn into a black date seed with six yellow spots. Before doing that, they shed their skin and attach it to themselves behind. The adult insects came forth at various times, some in January, some in February, some even in March (that is, of the following year). Their wings were white, adorned with black spots, the hind part of the body yellow, the two horns black, the six legs black and white. When they lay their yellow eggs, the little worms emerge within eight days, and as long as they are still small they feed only on the most tender part of the green leaves, not the whole leaf as the largest caterpillars do. If these large caterpillars, one of which is shown here crawling down the stem, are touched, they roll themselves up, but if they are squeezed they twist and turn violently. The cocoon is pictured above the date seed; the moth-bird can be seen on the opposite side with its eggs and tiny caterpillars.

Chapter 46: *Spilosoma lubricipeda* (white ermine moth) and
Amberboa moschata (sweet sultan)

This caterpillar feeds on a broad array of plants and is a good example of a
generalist herbivore. In this late chapter, we again have evidence that Merian
constructed general tenets based on her observations of many specific examples. Here, she summarizes what she has discerned about larval feeding from
her years of study and delineates some categories of larval food choices. She
was the earliest naturalist to write that some larvae required a specific food,
whereas others might have a strong food preference but could, when required,
survive on alternative host plants (or even cannibalism). A third group, which
included the white ermine moth caterpillars, had "a definite taste" for an assortment of food plants, preferring to eat a variety of things.

Another novel observation here was her finding that the smallest ermine
moth caterpillars fed only on the most tender leaf parts, whereas the larger
instars ate the entire leaf. Ecologists attribute this differential preference to
the ability of successive instars to ingest and successfully digest different parts
of plants. Plants use both chemical and physical defenses to reduce herbivory,
and these defenses – noxious compounds and tough structures, for example –
are not distributed equally throughout a plant.[1] Merian was the earliest to
publish extensive observations regarding the food choices of insects, a vital
area of ecology that determines the success of countless species of both insects
and plants. The significance of this field can be seen in the geometric increase
in the number of insect/plant studies published since the early 20th century;
interestingly, many of the earliest empirical studies had their roots in Germany
and Holland.[2]

Plate 46 shows tiny caterpillars emerging from eggs laid by the female, and
as in previous entries, Merian describes the defensive behavior of the larvae, something omitted from her study journal entry on this species. Once
again, she refers to the hairs of a caterpillar and the use of these in cocoon
construction. She illustrated this insect's life cycle on a relatively exotic plant,
the Sweet Sultan, introduced in Europe around 1630. She likely chose this
showy flower to spice up the images of plants offered in her book, making it
even more attractive to gardeners.

---

1   Hochuli, Dieter F. 2001. Insect herbivory and ontogeny: how do growth and development
    influence feeding behaviour, morphology and host use? *Austral Ecology*, 26(5): 565.
2   For a review, see Schoonhoven, L. M. 1996. After the Verschaffelt-Dethier era: the insect-plant
    field comes of age. *Entomologia Experimentalis et Applicata*, 80(1): 1–5.

PLATE 47   Merian, 1679. *Raupen*

XLVII. Plum fruit   *Fructus prunorum*

(95) Whether this caterpillar crawling along the bottom is not one of the most attractive and elegant types described up to now should soon be clear. For it has five large yellow hair bristles on its back and one more erect red tail of hair at its hind end. The rest of the caterpillar is a very pretty yellow, like the lovely color of egg yolk. When it stretches out, one can see, behind the head and between several segments, broad black bands like velvet. On either side of the body there are black spots. Below the head there are six small red claws on each side, and in the middle of the body are eight yellow legs, with two more at its hind end. It is by nature very easily alarmed, for as soon as it senses or feels the least thing, it curls itself together at once and lies there as if dead until everything is completely quiet again. At the end of August it entered its change and produced a white cocoon in which it changed into a brown date seed. I have shown one open (so that it can be seen, lying as if wrapped in a swaddling cloth), placed on a green leaf in the center. And since I had a number of them, (96) it happened that some of the adult insects emerged in November, but others in April of the following year. All were moths that fly only at night, for during the day they are quiet, as if unable to see well. Their color is white and an attractive gray, like the color of silver. They have two brown horns and six gray or silver-colored legs. The eggs they laid were the same color and were the size of little millet seeds, as can be seen above the caterpillar. One of the moths is resting on the leaf at the top. For garden lovers I shall add this further bit of advice: If one wishes to spoil or kill the eggs of all kinds of moth-birds or summer-birds, one must destroy them with smoke in the various places where they are found, since (as noted elsewhere) they are often bunched close together. But the small and larger caterpillars must be destroyed with fire where they rest, or at night when they lie on top of each other in large numbers. Although smoke is not sufficient for the caterpillars, it can certainly be enough to destroy their eggs.

Chapter 47: *Calliteara pudibunda* (pale tussock moth) and *Prunus domestica* (plum)

Merian viewed the larva of the pale tussock moth as one of the most attractive caterpillars that she raised, and it is perhaps fittingly called 'Merian's brush' by some entomologists today. As in other *Raupen* book entries, she used the text in Chapter 47 to describe the vibrant coloration of the insects as an aid to identification and to provide a detailed coloring guide for those who purchased an uncolored copy of the *Raupen* book. In her engraved plate, the larva is portrayed in side view, displaying its stripes and prominent red bristles. Her study journal image of the same caterpillar is very different and less successful; there she portrayed the caterpillar as it would appear from above, and the result is an amorphous yellow shape that is not easily identifiable (Figure 84). In the study journal, she painted the pupa with the cocoon dissected away just as it is shown in the plate, but for her book's image, she positioned the cocoon on a leaf and made note of this artificial arrangement in her text. In nature, pupation of most moths takes place out of sight in the soil, or as with this species, under leaves or fallen trees.

In addition to describing the defensive behavior of the tussock moth larvae and timing of adult flight, she again included her supposition that the adult moths cannot see well in the day. She concluded the chapter with a method for destroying her most beautiful caterpillar, by using fire when the larvae congregate at night. Merian again showed the eggs deposited on an upper leaf of the plant, and included the information that these as well as the eggs of other insects could be destroyed by exposing them to smoke. These 'pest control' tips provide further evidence that she saw gardeners as an important audience for her book.

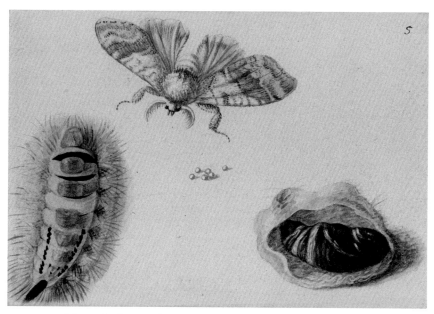

FIGURE 84   Merian. Study journal image 5. *Calliteara pudibunda* (pale tussock moth)

PLATE 48   Merian, 1679. *Raupen*

XLVIII. Ivy-leaved garden mallow

Malva, folio hederaceo

(97) Caterpillars such as the one crawling about here on the lowermost green leaf are brown and have a black head with a bit of yellow just behind it. They appear somewhat fat and ungainly. They have three claws on either side behind and below the head, and in the middle of the body four legs on each side and two more at the hind end. All of these, together with the claws, are ash-colored, as is the rest of the caterpillar. They take as their food or nourishment the attractive single, purple mallow, one of which can be seen here. Once they had attained their full size, they came to rest on these same green leaves and rolled them up, appearing to be spun up inside them. Then they changed into date seeds, curved like a half-moon, that were light blue but encircled with three narrow, bright-brown bands near the hind end. One of them is shown lying on a green leaf above the caterpillar. (98) The pretty summer-birds they produced emerged as late as November looking just like the one resting on the mallow and the other one in flight (for both are of the same type). Their outer sides appeared green and white, as did their body; the inner sides were the same colors, though somewhat more reddish. They were provided with six legs of the former color and two such horns. They fly rather slowly, but more actively and nimbly toward evening, which surprised me in these summer-birds. But some other caterpillars of the same type which I had collected did roll their leaves up in the same way and also remained alive without food in their rolled-up shelters into January of the following year, having changed into date seeds of the colors described above, and in June of that year the summer-birds I have described came forth just as clearly and beautifully.

Chapter 48: *Carcharodus alceae* (mallow skipper) and
Malva alcea (mallow)

The mallow skipper is another butterfly named for a preferred food plant of
the larvae. Merian did not name the insect, but as usual depicted the larva on
the correct host. In Plate 48, the caterpillar of the mallow skipper is shown
chewing a large hole in the mallow leaf. She details the positions of the caterpillar's prolegs on the body with particular care, and does the same in detailing
the delicate coloration of the pupa.

She describes the leaf-sheltering behavior of the mallow skipper caterpillar
as preceding pupation, but these caterpillars also shelter in rolled leaves as
earlier instars. Her account of the timing of pupation is a bit jumbled, possibly because the mallow skipper is bivoltine. She does indicate that adults may
appear two times in a year, both in November and in June. Merian correctly
describes the ability of this species of caterpillar to overwinter in a dormant
state, writing that they "remained alive without food," something that must
have amazed her, knowing how summer caterpillars that did not eat invariably
died. The subtle colors in the painted counterproof are very close to those in
the study image for this plate (Figure 85).

FIGURE 85 Merian. Study journal image 85 of *Carcharodus alceae* (mallow skipper). The caterpillar and pupa on the left are an example of an incomplete metamorphosis of a different species (unidentified).

PLATE 49   Merian, 1679. *Raupen*

XLIX. Purple carnation *Caryophyllus purpureus*

(99) This type of caterpillar is one that I have found at various times in July on cabbage and *Käsköhl*, otherwise known as cauliflower, also on sugar peas and carnations or grass flowers,[1] all of which they used for food, and with which I maintained them into the month of August. I have never seen these caterpillars eat during the day, but only at night; and the carnations, or grass flowers, were their favorite food. I have often come across them in the ground, too, where they remained all day but crawled out at night in search of their food. They are light brown with black stripes, and light yellow below. At the end of August they went into their change and turned into brown date seeds. The caterpillar and the date seed lie, one above the other, next to the grass flower stem in the lower center of the copperplate. Moreover, some caterpillars went to rest and shriveled up, like the one lying off to one side, and from which a maggot crawled out and immediately changed into a brown capsule, as I have shown lying between the maggot and the first caterpillar, (100) and from which after fourteen days a fly came out, like the one resting here on the green leaf. In this way another of the caterpillars also went to rest, and from it came four little maggots which changed immediately into as many little date seeds. Four such maggots and four small date seeds are pictured resting next to the large date seed. Although I have got such small worms a number of times, nothing more ever came of them, just as for three years in a row nothing more came from the large caterpillar described above, even though it had changed into a date seed each time. In the fourth year of my investigations, however, such a caterpillar once again went into its date seed in August, and in April of the following year a dark brown moth-bird with black stripes crawled out of it and can be seen here resting on the grass flowers at the top. It was completely quiet during the day, never moving, but at night it flew about like other moth-birds.

---

1   Carnations are supposedly called this because of their 'grass-like' foliage.

Chapter 49: *Mamestra brassicae* (cabbage moth) and *Dianthus caryophyllus* (carnation), Tachinid fly and other parasitoids

The cabbage moth was depicted on a garden flower rather than on the plant from which it now takes its common name. As Merian points out, the caterpillar of the cabbage moth eats plants other than the one from which it takes its common name, including 'grass flowers' (as she calls carnations). The species is a generalist herbivore and will indeed eat carnations, along with many other herbaceous plants.

Her statement that the caterpillar feeds only at night and hides during the day is also correct and is evidence of her diligence in observing insect behavior at all hours of the day and night. The caterpillar in her plate is posed on the ground and curled up as if resting, a position she could have observed in her captive insects. It took her four years to obtain a single adult moth from many efforts, reminding us once again that insect husbandry is a challenging endeavor requiring patience and experience. Parasitism was likely the cause for her difficulties with this species, as with some others. Here she reported a large fly (probably a Tachinid) resulting from a different type of metamorphosis that occurred in a "shriveled up" caterpillar; she also described a number of smaller maggots and pupae that did not hatch. Merian added one final behavioral observation concerning the adult cabbage moth, stating that it flew only at night "like other moth-birds."

PLATE 50   Merian, 1679. *Raupen*

L. Oak tree, with its fruit   *Quercus, cum fructu*

(101) Such caterpillars as the one crawling on the two oak leaves at the top have two yellow lines on each side; underneath they are striped yellow again and are otherwise brown in color, save that their head and the very hind part of their body, together with the claws and legs, are red. I have fed them on those same leaves into September, at which time they turned into red-brown date seeds, after they had first pushed off their skin and attached it in back, as can be seen lower on the twig. Then in December, quite lovely moth-birds came out, whose two outer wings [forewings] were brown and had some yellow flecks on them, but the inner ones [hind wings] were lighter brown. All of the rest was brown except the two eyes, which were black. Besides these, I have also had such caterpillars that were yellow and green at first (one is the smaller one crawling at the bottom); but after that, when they had become larger and had pushed off their skin, they became exactly like those described above in color. But since I had them longer, they pushed off their skin a second time and became, instead of the brown, a beautiful red color, like the one at the bottom of the page, and were also much larger than the one crawling at the top. But its date seed and insect were otherwise (102) the same as the one at the top. In the year now past, a caterpillar was brought to me that had been found on the grass and which I judged to have fallen from an oak. This one was much prettier in color, like a beautiful red rose. On that same day it made a white cocoon (which none of the others had done) and became a red-brown date seed and came out again as the above-mentioned, beautiful moth-bird, in form and color exactly like the one above.

Having now, by the grace of God, explained all of this, may it please the esteemed reader to know that it has all been done by me to the honor of God alone, in the hope that His glory and praise might shine forth more brightly and graciously among earthly-minded mankind, even from these things, in outward appearance poor and for many perhaps contemptible; considering that I would never have begun this laborious little work, much less been persuaded to have it appear in print, for any other reason, especially if it should be thought an unseemly, overweening ambition for me as a woman (who was obliged to compile it while managing her household). Thus, insofar as I shall see or hear God's praise increased thereby, I will henceforth (if I live longer and remain in good health), with God's help, let no occasion pass to add to it that which I have thought necessary to make known.

Chapter 50: *Ceramica pisi* (broom moth) and *Quercus robur* (pendunculate oak)

This final plate pictures the broom moth caterpillar, which feeds on a variety of plants well into September. Merian indicates the lateness of the season by including acorns on the beautifully figured oak branch, which shows caterpillar damage around the edges of the leaves. In study journal entry 15, the adult and pupa of the broom moth are shown with a dark brown caterpillar similar to the one on the leaves. She added two additional caterpillars to the plate; these were painted onto a nearby study painting with the number 16 inscribed but were not mentioned in her handwritten text for that entry. The earliest instar she describes as primarily green, and the last and largest as a "beautiful red color," although this is not obvious in the colored plate included here.

As previously mentioned, some color changes are due to diet and other factors. In some cases, as with the broom moth and other cryptic caterpillars, the later instars exhibit colors that more closely resemble the twigs on which they rest. The smaller, lighter colored instars are more likely to match the leaves on which they are almost constantly feeding. The broom moth typically overwinters in the pupal stage, and the imago emerges in the spring. However, she once again had some adults emerge in December, probably as a result of the warmth of her house.

As discussed earlier, Merian uses the end of this final chapter to reflect briefly on her motivation for the work, its challenges, and her plans for further research. Merian tells her readers that she did this work to honor God, and then states that she was "persuaded" to publish it, something she first mentions in her preface. These closing remarks again make clear her wish to avoid any appearance of unfeminine ambition, even as she reminds us that the work was "laborious." Her last sentence in the book is interesting on two levels. First, Merian indicates that if she hears that her book has elevated praise for God, she will "add to it," thereby laying the groundwork for the 1683 *Raupen* book. By this phrasing, she lets readers know that encouragement of her work would be welcome.

At the time that the first *Raupen* book was printed, she already had a great deal of research material for the 1683 *Raupen* (see p. 92 in this volume), and a second book must have been on her mind. The second part of this final sentence plays out decades later in information added to the last chapter of the Dutch edition of the 1679 caterpillar book. In the 1712 *Rupsen* edition, she alters Plate 50 to include an oak gall and describes her study of this puzzling structure in the revised text (see p. 114 in this volume). As with her other painstaking observations, she deemed the intriguing oak gall and the insects inside it "necessary to make known" to a new audience of readers in the 18th century. Thus, Merian's first caterpillar book concludes by setting the stage for her future work, through which her continuing curiosity about insects would bring new discoveries to a growing audience.

Index

of the most noteworthy items in caterpillars' transformation and food plants.

Eylein, so blau-grün. *eggs, blue-green.*   62

————, braun. ~ *brown.*   16, 18, 99

————, gelb-braun. ~ *yellow-brown.*   68

————, roth-braun. ~ *red-brown.*   90

————, gelb. ~ *yellow.*   92

————, gelb und schwartz. ~
yellow and black.   60

————, gold-gelb. ~ *golden yellow.*   32

————, holtz-gelb. ~ *wood yellow.*   44

————, leber-farb. ~ *liver-colored.*   22

————, weiß. ~ *white.*   46

————, weiß und meergrün. ~
white and sea-green.   36

F.

Fledermäuse, Nahm und Art gewisser
Sommer-Vögelein. *bats, name and
type of certain summer-birds.*   88

Farb der Motten- und Sommer-vögel.
color of moth- and summer-birds.   66

G.

Garten-fenchel. *garden fennel.*   77

Garten-Iris. *garden iris.*   71

Garten-müntze. *garden mint.*   79

Garten-pappelrose, klein. *garden mallow,
lesser.*   97

Garten-rittersporn. *garden larkspur.*   81

Gärtners-regel. *gardeners, advice for.*   37, 96

Gespinst, stark wie Pergament.
cocoon, hard as parchment.   20

————, wie gelbe Seiden. ~
like yellow silk.   68

————, wie Wolle. ~ *like wool.*   66

Gold-vögelein, sonst Zwifalter genennt.
*Gold-vögelein, otherwise called
Zwifalter.*   90

Gras, gemein, auf den Wiesen. *grass,
common, on meadows.*   65

Gunreben, samt ihrer Blüe. *ground ivy,
together with its blossom.*   69

H.

Haar-salbe, vom Alber- oder Pappelbaum.
hair cream, from the poplar tree.   85

Hagendorn, blüend. *hawthorn, in flower.*   41

Hanenfuß, süß. *spearwort, or buttercup,
sweet.*   13

————, wild. ~ *wild.*   31

Holder, blau. *elder, blue.*   7

Hyacinth, blau und Orientalisch.
hyacinth, blue, Oriental.   11

I.

Iris, im Garten. *iris, in the garden.*   71

J.

Johannesbeerlein blüend, klein und
roth. *currant in flower, small, red.*   29

————, groß und roth. ~ *large, red.*   75

————, weiß. ~ *white.*   59

K.

Kegelspiels Gestalt, im Samen.
bowling-pin arrangement, of eggs.   55

Kindes Gestalt, am Dattelkern zu sehen.
*form of an infant, visible in the date
seed.*   78

Kirschenblüe, schwartz und süß.
cherry in flower, black, sweet.   9

Kirschen, frühe und süsse. *cherries,
early, sweet.*   47

Klosterbeere. [*alt. name for*] *gooseberry.*   21

Köhlblüe, gelb. *kale, yellow.*   91

Kopfsalat, blüend. *head lettuce, in flower.*   87

Körbelkraut. *chervil.*   33

Kraut-vögelein. *cabbage birds.*   92

Kühblume. *dandelion.*   17

L.

Lattich-salat, blüend. *lettuce, in flower.*   87

Liljen, blau. *lilies* [*i.e., iris*], *blue.*   71

---

1 See Chapter 45, p. 92 and Chapter 22, p. 46 for context.

Index

V.

FINIS.

Translated Excerpts from Merian's Study Journal

The translations were made from the transcribed copy of Merian's study notes in a facsimile volume of her handwritten notes and painted studies.[1] The numbering system for the journal entries was her own, and numbers correspond with those on the study paintings. She refers to the position of the information from each entry in her published work with language such as "copperplate No. 1 in my first Raupenbuch." As mentioned earlier, the study notes for her 1679 and 1683 *Raupen* books were re-copied into the existing form after the works were published.

1   Because the silkworm's name is known to almost everyone, and it is the most useful and valuable of all the worms and caterpillars, I have placed its transformation here at the beginning:

Firstly, they eat the leaves of the white mulberry rather than those of the red. But because at the time when the silkworms are young no leaves have come out yet in our region, we give them lettuce leaves to eat in the meantime, often for as long as it takes them to reach the size of the one above the lowest cut-open date seed, near the large silkworm on the right. Their color is normally white, and finally, when it is as big as the just-noted large silkworm, it begins to spin, becomes somewhat yellowish, shrunken, and transparent, swings its head back and forth, and releases the silk from its mouth. At this time, it is put into a little paper cone, wide at the top and pointed at the bottom, called a *Scharmützel* in Nuremberg, but a *Dotte* in Frankfurt am Main. This is then placed away somewhere. Inside, it calmly spins a white, yellow, or greenish cocoon [*Ey*], a yellow example of which can be seen in the middle of the parchment, and inside it becomes a complete date seed, as also seen there. First, before spinning, it pushes its skin off several times, as can be seen above the large worm, behind the date seed. Then, when the moth is about to hatch, it chews its way through the date seed and cocoon, such a cocoon as seen at the very top, the date seed at the bottom; and its color is generally white, and it takes half a day until it gets full, dry wings or its complete moth form. The male is smaller and more delicate than the female and has a more slender body. As

---

1 Merian and Beer, *Leningrader Studienbuch.*

soon as they have their strength they mate, and that same day, or the day following and continuing for several days, they lay round, yellow eggs, like millet
seeds, and then die. On each little egg is a brownish dot, and if they are collapsed or appear like empty shells, nothing will come out, for they are spoiled.
In the sun or in a warm room they will crawl out sooner, that is, from the just-
noted dots. Then they are given lettuce right away. In cleaning [their contain-
ers], one must pay careful attention so as not to damage them with brusque
handling or the like; one must also not give them any wet leaves or any rotted
material, or they will deteriorate, become sick, and die. They molt when they
are barely 4 days old, and many die in the process. When a thunderstorm is
coming up and there is lightning, they must be covered; otherwise they will
contract jaundice or dropsy. They also die if given too much to eat. Their droppings are dark green.

Now, because after the description of the familiar so-called silkworm almost
all the transformations and changes of the worms, caterpillars, and maggots to
follow will be more readily understood, I made this noble transformation myself as copperplate No. 1 for my first *Raupenbuch* and described it in the greatest detail (without mentioning its fine uses, suggested above; but as a model
for instruction).

I began this investigation, praise be to God, in Frankfurt in 1660 and offered
the first part, from Nuremberg, at the fall [book] fair in Frankfurt in the year
1679; the second part, while residing in Frankfurt, in the year 1683.

Maria Sibyla Merianin

Anno 1660: My first observation in Frankfurt am Main[2]
2   When this caterpillar is found in August, it spins itself up in September
and remains resting in its date seed through the winter, into March of the following year; then its bird [adult] comes forth and is seen flying in the spring in
our area. But when the caterpillar comes in May or June, it remains only two
weeks in its change, or date seed, and the same kind of bird comes forth. Now
these caterpillars push their skin off several times before they get their full size,
and then it crawls onto a wall and pushes off the husk or skin completely one
more time, and after that it hangs head down, attaching its body as firmly as
if it were glued on, and turns into a date seed. If one tears it off, nothing will
come out and it will wither. But in many date seeds, very small white worms
will sometimes grow, some 60 or 70 of them. They spin themselves up in as
many little capsules [*Eylein*], after which just as many little golden flies come

---

2 Merian inserted this dated statement just above entry 2, which begins on a separate sheet
   and not on the sheet with the silkworm entry.

out, as two little worms and three little flies can be seen here.[3] But from a good date seed a two-colored bird, or summer-bird, comes forth, like the one pictured (shown both perched and flying). Now when one has many caterpillars together, many white worms will grow in their droppings, one of which can be seen next to the date seed, and on the same day it will turn into a brown capsule (shown lying below); then, after 14 days, a common fly comes out of this capsule. The droppings, like two little black grains, lie between the end of the caterpillar and the brown capsule. (This transformation appears in my *Raupen*, Part I, No. 44. Their food is the little stinging nettle.)

4   These three types of caterpillars take all kinds of roses as their food and do much damage to them, because they are usually found in the hearts of the roses and hollow them out completely. They crawl up through the stem, and one can see them crawling through the hollowed-out stem, as if through a tube. I take it that they grow in there. When they have eaten their way out and you touch them, they let themselves down by a thin thread that comes out of their mouth, and back up again; and at the end of May, when they have attained their full size, they become brown date seeds and remain lying that way into the middle of June; then from each one comes a similar insect, and they have a very rapid flight. As can all be seen pictured on the third vellum, in three series, one [stage] above the other. In the first series, one can also be seen at rest above the one in flight. The first transformation series is in my first *Raupen* volume, No. 24. The second transformation series, No. 19, also in the first *Raupen* volume. The third, in that same volume, but No. 28.

11   Caterpillars like the one seen in the first parchment have their start in April. They move very slowly and like to eat the green leaves of the *Kloster-* or gooseberries. They eat vigorously and like to have water; for they are refreshed by it if they are regularly given a few drops. One must especially do this when they are in danger of drying out, otherwise they will die. They also eat grass and the leaves of plums and roses. In June they make a round cocoon that is yellowish and quite hard, in which they become a brown date seed; and in July, yellow moth-birds come forth that fly only at night. I placed this transformation in my *Raupen*, Part I, No. 10.

25   This crawling black caterpillar, which has long, black hair on its back but yellow on its underside and head, is a voracious eater and feeds on all sorts of flowers and herbs such as grass, deadnettle, stinging nettle, and

---

3   Only two flies are shown in the study journal painting and in the published plate.

the like; hyacinth, for example, a stem of which is pictured in my *Raupen*, Part I, plate 5. They also move very rapidly. Gardeners call them *Beermutter*. If one touches them, they roll themselves up in a ball and remain like that until they no longer feel anything. Then they gradually stretch out again and crawl away. (I have had small caterpillars of this kind that placed themselves on a wall or wooden surface in April and grew quite hard, as shown here. These remained that way for ten days, when from some of them a completely black fly emerged, but from others a fly with some yellow in the middle of its body; one can be seen below the shrunken, hardened caterpillar.) They make a partly gray cocoon in which they become a date seed, pushing off their skin several times beforehand. But if one peels away the topmost layer before they emerge, one will see how the moths are placed [inside] and how small their wings are, as shown here below the date seed. They remain fourteen days in their transformation, then the mottled adult moth (above) emerges and soon lays its eggs [*Samen*], seen here below the moth.

I made this transformation No. 5 in my *Raupen*, Part I.

38   This kind of caterpillar feeds on all kinds of trees and is also very partial to moisture and drinking: hence there were so many thousands, and of such great size, in the year 1679, with its alternating warm rains and bright sunshine all spring; thus they were refreshed and grew ever larger. For when it rains too heavily it lacerates the young caterpillars and washes away those that seek shelter on the underside of green leaves, so that they are completely destroyed in the runoff. The same thing happens to their eggs [*Samen*]. Likewise, if there is too long a period of sunshine, many of them also die before they have spun cocoons or been able to find other sources of moisture. They have done enormous damage not only to fruit trees but also to lindens and willows—in fact, all kinds of other trees. Their trunks were entirely covered from top to bottom, and their leaves often stripped so bare that they appeared as if frost-killed. But here is some useful information for the attentive gardener: As evening approaches, they all crawl together onto one branch where they have their cocoon or nest. There they can all be captured at once and destroyed. They transform, some in June and some in July, and usually turn directly into a brown date seed; some, however, spin a light cocoon and remain lying that way for fourteen days. Then a large moth emerges, white with brown markings, as can be seen on the parchment. After several days they lay their eggs [*Eyer*] from which, in April of the following year, come small caterpillars that are all black. They make a kind of coat around the eggs such that one will not even notice them; and they are thus well protected from the cold—as the winter in question was very cold, yet it did not harm them. I made this transformation No. 18 in my *Raupen*, Part I.

119a This large, golden-yellow-and-black caterpillar (which, while still small, looks like the smaller one next to it) I found in large numbers in 1677 in Altdorf,[4] where the Nuremberg University is, in the grass of the town dry-moat, where they were feeding on the clover and sorrel. I fed them with that every day from August into September, after which they made a structure out of their droppings and green leaves and concealed themselves inside it. But nothing more came of them; they were entirely spoiled. And although I got them a number of times in the very next year and tended them with great care and all kinds of things to help them on to their transformation, and they managed to live through the winter, they eventually wasted away or turned to slime.

119b Once, in early July, when I went up to my garden (next to the castle church, or Imperial Castle Chapel, in Nuremberg), both to see the flowers and to look for caterpillars, I found much green mud on the green leaves of the golden-yellow lilies. I wanted to find out where the mud came from, touched it with a little stick, thinking that perhaps the leaves were rotting, and there I found in the mud very many small, round, red little creatures, like small beetles, sitting very close with their heads together, and completely motionless even though I touched them hard. So I took some of them home, along with the leaves they were lying on, to observe what might come of them. They remained lying in the unclean mud, but when I looked at them again several days later, I found that they had changed in the same way as I found later in my garden: like the stubby red date seed shown just above the green leaf (opposite the little beetle caterpillar[5] sitting on the leaf). Now in late July, beetles came out like the one crawling at the top. It laid its eggs, likewise red, on the leaves of the lily, and arranged as if they were bowling pins. Once again, little creatures came out of them like the ones that had been covered with mud. I placed this transformation in my *Raupen*, Part II, No. 21.

212 On this green oak leaf are three round growths called oak medlars [*Eichenmispeln*]. When I cut these open on 7 July 1684, in Schwalbach [in the Taunus Hills], in the presence of Dr. Faberitzius, Palatine Physician-in-Ordinary,

---

4 Altdorf is 25 km East of Nuremberg; there was a university there from 1622 to 1809.
5 Merian's term here is *Räuplein*, even though this larva does not have the usual form of a caterpillar.

we found in each one a little round hole in the center of the fruit, in which lay a round seed [*Samen*] like the one here lying next to the hole. But after twelve days, we went looking for them again, and after we had cut them open we found that the little round holes had formed and divided into two parts, like small hearts, and in each was a little white worm,

This transformation is in my first *Raupen* volume, No. 50.[6]

In Amsterdam, in 1710, I again found this kind of oak medlar, in which were such small black flies as the one shown here.[7]

---

6 She added this mention of Chapter 50 in her "first *Raupen* volume" after she re-etched some of the original plates for reprinting in the 1712 *Rupsen* book. See the commentary on this on p. 114

7 Merian made two slightly different handwritten copies of this entry, both of which are preserved in the study journal at the Library of the Russian Academy of Sciences, St. Petersburg. They are similar enough that only one is translated here.

Bibliography

Manuscripts

Adriaen Coenensz. 1577–1580. *Vis booc.* The manuscript is in the Jacob Visser collection, National Library of the Netherlands. Shelfmark 78 E 54.

Hoefnagel, Joris. *Animalia Rationalia et Insecta (Ignis).* Volume 1 of a manuscript in the National Gallery of Art. Accession number 1987.20.5.

Merian, Maria Sibylla. Letter written to Clara Regina Imhoff in Nuremberg, dated 25 July 1682. Germanisches Nationalmuseum Nürnberg, Imhoff Archive.

Merian, Maria Sibylla. Letter written to Clara Regina (Imhoff) Scheurling in Nuremberg, dated 29 August 1697. Autograph Number 167, Stadtbibliothek Nürnberg.

Merian, Maria Sibylla. Letter to James Petiver in London, dated 27 April 1705. Folio 70, Sloane 4064. British Library.

Online

Audusseau, Hélène, Gundula Kolb, and Niklas Janz. 2015. Plant fertilization interacts with life history: variation in stoichiometry and performance in nettle-feeding butterflies. *PLOS ONE, 10*(5): 9. https://www.ncbi.nlm.nih.gov/pmc/articles/PMC4416804/ Accessed 6 March 2020.

Mulder, Hans. 2014. Merian puzzles. Some remarks on publication dates and a portrait by Johannes Thopas, in: *Proceedings of the symposium "Exploring M. S. Merian,"* Amsterdam, 2014 on the website of *The Maria Sibylla Merian Society.* https://www.themariasibyllameriansociety.humanities.uva.nl/research/essays2014/ Accessed 26 February 2020.

Noor, Mohamed A. F., Robin S. Parnell, and Bruce S. Grant. 2008. A reversible color polyphenism in American peppered moth (*Biston betularia cognataria*) caterpillars. *PLOS ONE, 3*(9): 1–5. https://www.ncbi.nlm.nih.gov/pmc/articles/PMC2518955/ Accessed 6 March 2020.

In Print

Abbot, John and James E. Smith. 1797. *The natural history of the rarer lepidopterous insects of Georgia: including their systematic characters, the particulars of their several metamorphoses, and the plants on which they feed. Collected from the observation of Mr. John Abbot, many years resident in that country.* London: Printed by T. Bensley, for J. Edwards.

Ackerman, James S. 1985. The involvement of artists in Renaissance science. In *Science and the Arts in the Renaissance*. John W. Shirley and F. David Hoeniger, Eds. Washington, DC: Folger Shakespeare Library: 94–129.

Albin, Eleazar. 1720. *A Natural History of English Insects: Illustrated with a Hundred Copper Plates, Curiously Engraven from the Life*. London: Published by the author.

Aldrovandi, Ulisse. 1599. *Vlyssis Aldrovandi ... Ornithologiae, hoc est de Avibus Historiae, Libri XII*. Bononiae: Franciscum de Franciscis Senensem.

Aldrovandi, Ulisse, Gio Battista Bellagamba, Giovanni Luigi Valesio Battista, and Francesco Maria Della Rovere. 1602. *De Animalibus Insectis Libri septem: Cum singulorum Iconibus ad Vivum expressis. Libri VII. Bonon: Apud Ioan: Bapt: Bellagambam cum consensu superiorum*.

Ashworth, William B. 1996. Emblematic natural history of the Renaissance. In *Cultures of Natural History*. Edited by Nicholas Jardine, James A. Secord and Emma C. Spary. Cambridge: Cambridge University Press.

Balbín, Bohuslav. 1679. *Miscellanea Historica Regni Bohemiae*. Pragæ: Typis Georgii Czernoch.

Baldassarri, Fabrizio. 2017. Introduction: gardens as laboratories. A history of botanical sciences. *Journal of Early Modern Studies*, 6(1): 9–19.

Balevski, Nikolai A. 1999. *Catalogue of the Braconid parasitoids (Hymenoptera: Braconidae) Isolated from Various Phytophagous Insect Hosts in Bulgaria*. Sofia: Pensoft: 48–50.

Ball, Philip. 2012. Nature's color tricks. *Scientific American*, 306(5): 74–9.

Bass, Marisa A. 2017. Mimetic obscurity in Joris Hoefnagel's Four Elements. In *Emblems and the Natural World*. Edited by Karl A. E. Enenkel and Pamela J. Smith. Leiden: Brill.

Bass, Marisa Anne. *Insect Artifice: Nature and Art in the Dutch Revolt*. Princeton: Princeton University Press, 2019.

Blair, Ann and Kaspar von Greyerz (Eds.) 2020. *Physico-theology: Religion and Science in Europe, 1650–1750*. Baltimore: Johns Hopkins University Press.

Blom, Franciscus Joannes Maria. 1981. *Christoph & Andreas Arnold and England. The Travels and Book Collections of Two Seventeenth-Century Nurembergers*. Ph.D. thesis, Catholic University Nijmegen.

Blunt, Wilfrid and William Thomas Stearn. 1950. *The Art of Botanical Illustration: An Illustrated History*. North Chelmsford, MA: Courier Corporation.

Bodenheimer, Frederick S. 1928. *Materialien zur Geschichte der Entomologie bis Linné*. Berlin: W. Junk: 401–7.

Bowen, Karen L. and Dirk Imhof. 2003. Reputation and wage: the case of engravers who worked for the Plantin Moretus Press. *Simiolus: Netherlands Quarterly for the History of Art*, 30: 161–95.

Bredius, A. 1915. *Künstler-Inventare: Urkunden zur Geschichte der Holländischen Kunst des 16ten, 17ten und 18ten Jahrhunderts*. Den Haag: M. Nijhoff.

Bry, Johann Theodor de and Matthaeus Merian. 1641. *Florilegium Renovatum et Auctum: variorum maximeque rariorum Germinum, Florum ac Plantarum*. Frankfurt: Matthaeus Merian.

Buckton, George B. 1883. *Monograph of the British Aphides*. London: Ray Society, Volume 4.

Buckton, George. 1895. *Natural History of Eristalis tenax or the Drone-fly*. London and New York: Macmillan and Company.

Camerarius, Joachim. 1596/1604. *Symbolorum & Emblematum ... centuria vna* [-quarta], Nuremberg: [part 1] for Johann Hofmann and Hubert Camoxius, 1590 [1593]: [part 2] P. Kaufmann, 1595: [part 3] ibid. 1596 [1597]: [part 4], 1604.

Capinera, John L. 2008. *Encyclopedia of Entomology*. Volume 4. Berlin: Springer Science & Business Media.

Christie's London. 2015. *Old Master and British Paintings*. London: Christie's. Auction catalog.

Colditz-Heusl, Silke, Margot Lölhöffel, Theo Noll, and Werner Schultheiss. 2017. *Katalog zur Ausstellung: Johann Andreas Graff, Pionier Nürnberger Stadtansichten im Kunstkabinett der Stadtbibliothek Nürnberg*. Nuremberg: Association of Friends of the Nuremberg Cultural History Museum.

Cooper, Alix. 2007. *Inventing the Indigenous. Local Knowledge and Natural History in Early Modern Europe*. Cambridge: Cambridge University Press.

Darwin, Charles. 1993. *The Correspondence of Charles Darwin*. Volume 8 (1860). Cambridge: Cambridge University Press.

Davis, Natalie Zemon. 1997. *Women on the Margins: Three Seventeenth-Century Lives*. Cambridge, MA: Harvard University Press.

Dicke, Marcel. 2000. Insects in western art. *American Entomologist, 46*(4): 228–36.

Dickenson, Victoria. 1998. *Drawn from Life: Science and Art in the Portrayal of the New World*. Toronto: University of Toronto Press.

Disney, R. H. L. 1994. Larvae. In *Scuttle Flies: The Phoridae*. New York: Springer: 34.

Doppelmayr, Johann Gabriel. 1730. *Historische Nachricht von den Nürnbergischen Mathematicis und Künstlern*. Nuremberg.

Egerton, Frank N. 2012. *Roots of Ecology: Antiquity to Haeckel*. Oakland, CA: University of California Press: 194–5.

Egmond, Florike. 2017. *Eye for Detail: Images of Plants and Animals in Art and Science 1500–1630*. London: Reaktion Books Limited.

Enenkel, Karl A. E. 2014. The species and beyond: classification and the place of hybrids in Early Modern zoology. *Zoology in Early Modern Culture: Intersections of Science, Theology, Philology, and Political and Religious Education*. Leiden: Brill: 55–148.

Erasmus, Desiderius. 1508. *Erasmi Roterodami Adagiorum Chiliades Tres*. Venice: Aldus Manutius.

Essig, Edward O. 1936. A sketch history of entomology. *Osiris*, 2: 80–123.

Etheridge, Kay. 2011. Maria Sibylla Merian and the metamorphosis of natural history. *Endeavour, 35*: 16–22.

Etheridge, Kay and Florence Pieters. 2015. Maria Sibylla Merian (1647–1717): Pioneering naturalist, artist, and inspiration for Catesby. In *The Curious Mr. Catesby: A 'Truly Ingenious' Naturalist Explores New Worlds*. Edited by E. Charles Nelson and David Elliot. Athens, GA: University of Georgia Press: 39–56.

Etheridge, Kay. 2016. The biology of *Metamorphosis Insectorum Surinamensium*. In: Merian, Van Delft, and Mulder, *Merian Metamorphosis*: 29–39.

Fludd, Robert. 1617. *Utriusque Cosmi Maioris Scilicet et Minoris Metaphysica, Physica atqve Technica Historia*. Oppenheim: Theodor de Bry. Volume 1.

Frisch, Johann Leonhard. 1720. *Beschreibung von allerley Insecten in Teutsch-Land: nebst nützlichen Anmerckungen und nöthigen Abbildungen von diesem kriechenden und fliegenden inländischen Gewürme*. Theil 1. Berlin: Nicolai.

Fürst, Rosina Helena. 1650. *Neues Modelbuch von unterschiedlicher Art der Blumen*. Nürnberg: Paul Fürst.

Gessner, Conrad. 1551–1558. *Historia Animalium*. Five volumes. Zurich: C. Froschauer.

Ghiradella, Helen. 1991. Light and color on the wing: structural colors in butterflies and moths. *Applied Optics, 30*(24): 3492–500.

Glardon, Philippe. 2007. The relationship between text and illustration in mid-sixteenth-century natural history treatises. In *A Cultural History of Animals in the Renaissance*. Edited by Bruce Boehrer. Oxford: Berg: 117–45.

Goedaert, Johannes and Johannes de Mey. 1660–1669. *Metamorphosis Naturalis*. Middelburgh: J. Fierens.

Goedaert, Johannes and Martin Lister. 1682. *Johannes Goedart of Insects*. York: printed by John White for Martin Lister.

Goedaert, Johannes and Martin Lister. 1685. *Johannes Goedartius de Insectis, in Methodum Redactus, cum Notularum Additione. Operâ M. Lister*. London: S. Smith.

Goldgar, Anna. 2008. *Tulipmania: Money, Honor, and Knowledge in the Dutch Golden Age*. Chicago: University of Chicago Press.

Grieb, Manfred H. (Ed.) 2011/2007. *Nürnberger Künstlerlexikon: Bildende Künstler, Kunsthandwerker, Gelehrte, Sammler, Kulturschaffende und Mäzene vom 12. bis zur Mitte des 20. Jahrhunderts*. Munich: K. G. Saur.

Gross, Paul. 1993. Insect behavioral and morphological defenses against parasitoids. *Annual Review of Entomology, 38*(1): 251–73.

Guilding, Lansdown. 1834. Observations on the work of Maria Sibylla Merian on the insects etc. of Surinam. *Magazine of Natural History and Journal of Zoology, Botany, Mineralogy, Geology and Meteorology*, 7: 355–75.

Harkness, Deborah E. 2009. Elizabethan London's naturalists and the work of John White. In *European Visions – American Voices*. Edited by Kim Sloan. London: British Museum: 44–50.

Harris, Moses. 1766. *The Aurelian. A Natural History of English Moths and Butterflies, Together with the Plants on Which They Feed. Also a Faithful Account of Their Respective Changes, Their Usual Haunts When in the Winged State, and Their Standard Names as Established by the Society of Aurelians.* London: published by the author and sold by J. Edwards.

Hochuli, Dieter F. 2001. Insect herbivory and ontogeny: how do growth and development influence feeding behaviour, morphology and host use? *Austral Ecology*, 26(5): 565.

Hoefnagel, Joris and Jacob Hoefnagel. 1592. *Archetypa Studiaque Patris Georgii Hoefnagelii.* Frankfurt: Jacob Hoefnagel.

Hoefnagel, Joris, Jacob Hoefnagel, Paul Fürst, and Albert Pfähler. Circa 1665. *Archetypa Studiaque Patris Georgii Hoefnagelii.* Nuremberg: Paul Fürst.

Hooke, Robert. 1665. *Micrographia, or, Some Physiological Descriptions of Minute Bodies Made by Magnifying Glasses with Observations and Inquiries Thereupon* / by R. Hooke ... London: Printed by Jo. Martyn and James Allestry, printers to the Royal Society.

Hünniger, Dominik. 2018. Nets, labels and boards: materiality and natural history practices in continental European manuals on insect collecting 1688–1776. In *Naturalists in the Field: Collecting, Recording and Preserving the Natural World from the Fifteenth to the Twenty-First Century*. Edited by Arthur MacGregor. Emergence of Natural History, Volume 2. Leiden: Brill.

Ikonen, Arsi. 2002. Preferences of six leaf beetle species among qualitatively different leaf age classes of three salicaceous host species. *Chemoecology*, 12(1): 23–8.

Jonston, Jan, Matthaeus Merian, and Caspar Merian. 1653. *Historiae Naturalis de Insectis, Libri III*. Frankfurt-am-Main: Merian.

Jonston, Jan and Matthias Graus. 1660. *I. Jonstons naeukeurige Beschryving van de Natuur der vier-voetige Dieren, Vissen en bloedlooze Waterdieren, Vogelen, Kronkeldieren, Slangen en Draken*. Amsterdam: Jan Jacobsz Schipper.

Jorink, Eric. 2010. *Reading the Book of Nature in the Dutch Golden Age, 1575–1715*. Leiden: Brill.

Jürgensen, Renate. 2002. *Bibliotheca norica: Patrizier und Gelehrtenbibliotheken in Nürnberg zwischen Mittelalter und Aufklärung*, Vol. 1. Wiesbaden: Harrassowitz.

Kinukawa, Tomomi. 2011. Natural history as entrepreneurship: Maria Sibylla Merian's correspondence with J. G. Volkamer II and James Petiver. *Archives of Natural History*, 38(2): 313–27.

Knight, Alan L., Mark Weiss, and Tom Weissling. 1994. Diurnal patterns of adult activity of four orchard pests (Lepidoptera: Tortricidae) measured by timing trap and actograph. *J. Agric. Entomology* 11(2): 125–136.

Koreny, Fritz. 1988. *Albrecht Dürer and the Animal and Plant Studies of the Renaissance.* Boston: Little, Brown and Company.

Kusukawa, Sachiko. 2010. The sources of Gessner's pictures for the *Historia animalium. Annals of Science, 67*(3): 303–28.

Kusukawa, Sachiko. 2011. Patron's review. The role of images in the development of Renaissance natural history. *Archives of Natural History, 38*(2): 189–213.

Kusukawa, Sachiko. 2012. *Picturing the Book of Nature: Image, Text, and Argument in Sixteenth-Century Human Anatomy and Medical Botany.* Chicago: University of Chicago Press.

Kusukawa, Sachiko. 2019. *Ad vivum*: Images and knowledge of nature in Early Modern Europe. In *Ad vivum?* Edited by Thomas Wolf, Joanna Woodall, and Claus Zittel. Leiden: Brill: 89–211.

Leith-Ross, Prudence and Henrietta McBurney. 2000. *The Florilegium of Alexander Marshal: In the Collection of Her Majesty the Queen at Windsor Castle.* London: Royal Collection.

Leonard, David. E. 1972. Survival in a gypsy moth population exposed to low winter temperatures. *Environmental Entomology, 1*(5): 549–54.

Lesser, Friedrich C. 1738. *Insecto-Theologia.* Frankfurt: Michael Blochberger.

Leßmann, Sabina. 1993. Susanna Maria Von Sandrart: women artists in 17th-century Nürnberg. *Woman's Art Journal, 14*(1): 10–4.

Linnaeus, Carl. 1751. *Philosophia Botanica in qua explicantur Fundamenta Botanica cum Definitionibus Partium, Exemplis Terminorum, Observationibus Rariorum, adjectis Figuris Aeneis.* Stockholm: Godofr. Kiesewette.

Lölhöffel, Margot. 2015. Maria Sibylla Merianin und Johann Andreas Graff: Gemeinsames und Trennendes, Erster Teil. *Nürnberger Altstadtberichte, 40*: 36–76.

Lölhöffel, Margot. 2016. Maria Sibylla Merianin und Johann Andreas Graff. Gemeinsames und Trennendes, Zweiter Teil. *Nürnberger Altstadtberichte, 41*: 64–116.

Ludwig, Heidrun. 1998. The *Raupenbuch*. A popular natural history. In Merian and Wettengl. 1998. *Maria Sibylla Merian:* 52–67.

MacGregor, Arthur. 2018. *Naturalists in the Field. Collecting, Recording and Preserving the Natural World from the Fifteenth to the Twenty-First Century.* Emergence of Natural History, Volume 2. Leiden: Brill.

Malpighi, Marcello. 1669. *Marcelli Malpighii Philosophi & Medici Bononiensis Dissertatio Epistolica de Bombyce.* London: Joannem Martyn & Jacobum Allestry.

Margócsy, Daniel. 2014. *Commercial Visions: Science, Trade, and Visual Culture in the Dutch Golden Age.* Chicago: University of Chicago Press.

Martinsen, G. D., K. D. Floate, A. M. Waltz, G. M. Wimp, and T. G. Whitham. 2000. Positive interactions between leaf rollers and other arthropods enhance biodiversity hybrid cottonwoods. *Oecologia, 123*(1): 82–9.

Mattioli, Pietro A. 1568. *I Discorsi di M. nelli sei libri di Pedacio Discodoride Anazarbeo della materia medicinale.* Venezia: Vincenzo Valgrisi.

Mattioli, Pietro Andrea, Joachim Camerarius, and Francesco Calzolari. 1586. *De Plantis Epitome vtilissima*. Frankfurt: Feyerabend.

Merian, Maria Sibylla. 1675. *Florum Fasciculus Primus*. Nuremberg: Johann Andreas Graff.

Merian, Maria Sibylla. 1679. *Der Raupen wunderbare Verwandlung und sonderbare Blumen-nahrung*. Nuremberg: M. S. Merian.

Merian, Maria Sibylla. 1680. *Neues Blumenbuch*. Nuremberg: Johann Andreas Graff.

Merian, Maria Sibylla. 1675–1680. *Florum Fasciculus Primus [Alter–Tertius]*. Nuremberg: Johann Andreas Graff.

Merian, Maria Sibylla. 1683. *Der Raupen wunderbare Verwandlung und sonderbare Blumen-nahrung ... Andrer Theil*. Frankfurt and Leipzig: M. S. Merian.

Merian, Maria Sibylla. 1705. *Metamorphosis Insectorum Surinamensium*. Amsterdam: G. Valck.

Merian, Maria Sibylla. [1712]. *Der rupsen begin, voedzel en wonderbaare verandering*. Volumes 1 and 2. Amsterdam: Maria Sibylla Merian.

Merian, Maria Sibylla. 1717. *Der Rupsen Begin, Voedzel en wonderbaare Verandering: waar in de Oorspronk, Spys en Gestaltverwisseling: als ook de Tyd, Plaats en Eigenschappen der Rupsen, Wormen, Kapellen, Uiltjes, Vliegen, en andere diergelyke Bloedelooze Beesjes vertoond word*. Volumes 1–3. Amsterdam: Published by the author.

Merian, Maria Sibylla. 1718. *Erucarum Ortus, Alimentum et paradoxa Metamorphosis ...* Amsterdam: Johannes Oosterwijk.

Merian, Maria Sibylla. 1719. *Dissertatio de generatione et metamorphosibus insectorum surinamensium*. Amsterdam: J. Oosterwijk.

Merian, Maria Sibylla and Wolfgang Dietrich Beer. 1976. *Maria Sibylla Merian: Schmetterlinge, Käfer und andere Insekten. Leningrader Studienbuch* ["Book of Notes and Studies"]. Luzern: Reich.

Merian, Maria Sibylla, Pierre-Joseph Buc'hoz, Jean Rousset, and Jean Marret. 1771. *Histoire générale des Insectes de Surinam et de toute l'Europe*. Paris: Chez L. C. Desnos.

Merian, Maria Sibylla, Thomas Bürger, and Marina Heilmeyer. 1999. *Neues Blumenbuch* = New Book of Flowers. Munich: Prestel.

Merian, Maria Sibylla and Jean Marret. 1730. *Histoire des Insectes de l'Europe, dessinée d'apres nature & expliquée par Marie Sibille Merian*. Amsterdam: Chez Jean Frederic Bernard.

Merian, Maria Sibylla and Jean Marret. 1730. *De Europische Insecten: naauwkeurig onderzogt, na't leven geschildert, en in print gebragt*. Amsterdam: J. F. Bernard.

Merian, Maria Sibylla, Elizabeth Rücker, and William T. Stearn. 1980–1982. *Metamorphosis Insectorum Surinamensium*. London: Prion

Merian, Maria Sibylla, Marieke van Delft, and Hans Mulder. 2016. *Maria Sibylla Merian. Metamorphosis insectorum Surinamensium = Verandering der Surinaamsche insecten = Transformation of the Surinamese insects: 1705*. Tiel, Belgium/The Hague, Netherlands: Lannoo.

Merian, Maria Sibylla and Kurt Wettengl. 1998. *Maria Sibylla Merian, 1647–1717: Artist and Naturalist.* Edited by K. Wettengl. Ostfildern-Ruit: Verlag Gerd Hatje.

Merian, Matthaeus. 1646. *Der fruchtbringenden Gesellschaft Nahmen, Vorhaben, Gemählde und Wörter.* Frankfurt: Merian.

Miall, Louis Compton. 1912. *The Early Naturalists, Their Lives and Work, 1530–1789.* London: Macmillan.

Moffet, Thomas. 1599. *The Silkewormes and their Flies.* London: By Valentine Simmes for Nicholas Ling.

Moffet, Thomas, Edward Wotton, Conrad Gessner, Thomas Penny, and Théodore Turquet de Mayerne. 1634. *Insectorum sive Minimorum Animalium Theatrum.* London: Thomas Cotes.

Müller, Kurt and Gisela Krönert. 1969. *Leben und Werk von G. W. Leibniz: eine Chronik.* Frankfurt: Vittorio Klostermann.

Neri, Janice. 2008. Between observation and image: representations of insects in Robert Hooke's Micrographia. In *Art of Natural History.* Edited by Theresa O'Malley and Amy Meyers. Washington DC: National Gallery of Art: 83–107.

Neri, Janice. 2011. *The Insect and the Image: Visualizing Nature in Early Modern Europe, 1500–1700.* Minneapolis: University of Minnesota Press.

Nickelsen, Kärin. 2018. Images and nature. In *Worlds of Natural History,* Edited by Helen Anne Curry, Nicholas Jardine, James A. Secord and Emma C. Spary. Cambridge: Cambridge University Press.

Ogilvie, Brian. 2005. Natural history, ethics, and physico-theology. In *Historia: Empiricism and Erudition in Early Modern Europe.* Edited by Gianna Pomata and Nancy G. Siraisi. Cambridge, MA: MIT Press: 75–103.

Ogilvie, Brian W. 2006. *The Science of Describing: Natural History in Renaissance Europe.* Chicago: University of Chicago Press.

Ogilvie, Brian W. 2008. Nature's Bible: insects in seventeenth century European art and science. *Tidsskrift for Kulturforskning, 7*(3): 5–21.

Ogilvie, Brian. 2015. Maria Sibylla Merian et la mouche porte-lanterne du Surinam. Naissance et disparition d'un fait scientifique. In *Les savoirs-mondes. Mobilités et circulation des savoirs depuis le Moyen Âge,* ed. Pilar González-Bernaldo and Liliane Hilaire-Peréz. Rennes: Presses Universitaires de Rennes: 147–57.

Ogilvie, Brian W. 2016. Willughby on insects. In *Virtuoso by Nature: The Scientific Worlds of Francis Willughby FRS (1635–1672).* Edited by Tim Birkhead. Emergence of Natural History, Volume 1. Leiden: Brill: 335–59.

Opitz, Donald L., Staffan Bergwik, and Brigit Van Tiggelen. 2016. *Domesticity in the Making of Modern Science.* New York: Springer.

Pettegree, Andrew. 2007. *The French Book and the European Book World.* Leiden: Brill.

Pick, Cecilia. 2004. *Rhetoric of the Author Presentation: The Case of Maria Sibylla Merian.* Doctoral Dissertation, University of Texas, Austin.

Pierce, Naomi. E. 1995. Predatory and parasitic Lepidoptera: carnivores living on plants. *Journal of the Lepidopterists Society, 49*(4): 412–53.

Pike, Nathan, Denis Richard, William Foster, and L. Mahadevan. 2002. How aphids lose their marbles. *Proceedings of the Royal Society of London B: Biological Sciences,* 269(1497): 1211–5.

Poulton, Edward B. 1903. Experiments in 1893, 1894, and 1896 upon the colour-relation between lepidopterous larvæ and their surroundings, and especially the effect of lichen-covered bark upon *Odontopera bidentata, Gastropacha quercifolia,* etc. *Transactions of the Royal Entomological Society of London, 51*(3): 311–74.

Rabel, Daniel. 1622. *Theatrum Florae in quo ex toto Orbe Selecti Mirabiles Venustiores ac praecipui Flores tanquam ab ipsus Deae sinu proferuntur Guillelmus Theodorus pinxit.* Paris: N. de Mathonière.

Rackham, Harris. 1940. Pliny. *Natural History,* Volume III, Book 11. Cambridge, MA: Harvard University Press.

Ray, John. 1710. *Historia insectorum opus posthumum jussu Regiae Societatis Londinensis editum.* London: Impensis A. & J. Churchill.

Ray, John, William Derham, and E. Lankester, 1848. *The Correspondence of John Ray.* London: Printed for the Royal Society.

Réaumur, René-Antoine Ferchault de. 1734–1742. *Mémoires pour servir à L'histoire des Insectes.* Six volumes. Paris: De l'Imprimerie Royale

Reitsma, Ella and Sandrine A. Ulenberg. 2008. *Maria Sibylla Merian and Daughters: Women of Art and Science.* Amsterdam: Rembrandt House; Los Angeles: J. Paul Getty Museum.

Renwick, J. A. A., and Chew, F. S. 1994. Oviposition behavior in Lepidoptera. *Annual Review of Entomology, 39*(1): 377–400.

Robert, Nicolas. 1660. *Variae ac multiformes florum species appressae ad vivum et aeneis tabulis incisae.* Paris: F. Poilly.

Roos, Anna Marie. 2018. *Martin Lister and his Remarkable Daughters: Art and Science in the Seventeenth Century.* Oxford: Bodleian Library Press.

Rösel von Rosenhof, August Johann. 1746–61. *Der Monatlich-Herausgegebenen Insecten-Belustigung.* Four Volumes. Nuremberg: Johann Joseph Fleischmann.

Roth, Michael, Magdalena Bushart, and M. Sonnabend. 2017. *Maria Sibylla Merian und die Tradition des Blumenbildes.* Munich: Hirmer Verlag.

Sandrart, Joachim. 1675/1679/1680. *Teutsche Academie der Bau-, Bild- und Mahlerey-Künste,* Nürnberg.

Sandre, Siiri-Li, Ants Kaasik, Ute Eulitz, and Toomas Tammaru. 2013. Phenotypic plasticity in a generalist insect herbivore with the combined use of direct and indirect cues. *Oikos, 122*(11): 1626–35.

Schmidt-Loske, Katharina. 2007. *Die Tierwelt der Maria Sibylla Merian (1647–1717): Arten, Beschreibungen und Illustrationen.* Marburg: Basilisken-Presse.

Schoonhoven, L. M. 1996. After the Verschaffelt-Dethier era: the insect-plant field comes of age. *Entomologia Experimentalis et Applicata*, *80*(1): 1–5.

Sepp, Jan Christiaan. 1762–1860. *Beschouwing der Wonderen Gods: in de minstgeachte Schepzelen of Nederlandsche Insecten ... naar 't Leven naauwkeurig getekent, in 't Koper gebracht en gekleurd*. Amsterdam: J. C. Sepp.

Smith, Pamela. H. 2004. *The Body of the Artisan: Art and Experience in the Scientific Revolution*. Chicago: University of Chicago Press.

Stearn, William T. 1978. *The Wondrous Transformation of Caterpillars: Fifty Engravings Selected from 'Erucarum Ortus'* (1718). London: Scolar Press.

Stijnman, Ad. 2012. *Engraving and Etching 1400–2000: a history of the development of manual Intaglio printmaking processes*. Houten, Netherlands: Archetype Publications in association with Hes en De Graaf Publishers.

Sugiura, Shinji and Kazuo Yamazaki. 2006. The role of silk threads as lifelines for caterpillars: pattern and significance of lifeline-climbing behaviour. *Ecological Entomology*, *31*(1): 52–7.

Swammerdam, Johannes. 1669. *Historia Generalis Insectorum, ofte Algemeene Verhandeling van de Bloedeloose Dierkens*. Utrecht: M. van Dreunen.

Taegert, Werner. 1998. Man's life is like a flower – Maria Sibylla Merian's watercolors for *Alba Amicorum*. In Merian and Wettengl. 1998. *Maria Sibylla Merian*: 88–93.

Terrall, Mary. 2014. *Catching Nature in the Act: Réaumur and the Practice of Natural History in the Eighteenth Century*. Chicago: University of Chicago Press.

Terrall, Mary. 2018. Experimental natural history. In *Worlds of Natural History*. Edited by Helen Anne Curry, Nicholas Jardine, James A. Secord, and Emma C. Spary. Cambridge: Cambridge University Press.

Thanos, Costas A. 1994. Aristotle and Theophrastus on plant-animal interactions. In *Plant-Animal Interactions in Mediterranean-type Ecosystems*. New York: Springer: 3–11.

Todd, Kim. 2007. *Chrysalis: Maria Sibylla Merian and the Secrets of Metamorphosis*. Orlando: Harcourt.

Topper, David R. 1996. *Towards an Epistemology of Scientific Illustration*. Toronto: University of Toronto Press.

Topsell, Edward, Conrad Gessner, Thomas Moffet, and John Rowland, M.D. 1658. *The History of Four-footed Beasts and Serpents ...* London: Printed by E. Cotes, for G. Sawbridge.

Trepp, Anne-Charlott. 2005. Nature as religious practice in seventeenth-century Germany. In *Religious Values and the Rise of Science in Europe*. Edited by John Brooke and Ekmeleddin İhsanoğlu. Istanbul: Research Center for Islamic History, Art and Culture: 81–110.

Uffenbach, Zacharias Conrad von. 1753. *Herr Zacharias Conrad von Uffenbach Merckwürdige Reise durch Niedersachsen Holland und Engelland.* Ulm: auf Kosten Johann Friedrich Gaum.

Valiant, Sharon. 1993. Maria Sibylla Merian: Recovering an eighteenth-century legend. *Eighteenth-Century Studies, 26*(3): 467–97.

Van Groesen, Michiel. 2008. *The Representations of the Overseas World in the De Bry Collection of Voyages (1590–1634).* Leiden: Brill.

Van Lenteren, Joop C. and Charles H. Godfray. 2005. European science in the Enlightenment and the discovery of the insect parasitoid life cycle in the Netherlands and Great Britain. *Biological Control, 32*(1): 12–24.

Van de Plas, Joos. 2019. *Het gestolen kijken – Stolen Observations.* Helvoirt: Published by the author.

Van de Roemer, Bert. 2016. Merian's network of collector-naturalists. In Merian, Van Delft, and Mulder, *Merian Metamorphosis*: 19–28.

Van Veen, Henk Th. and Andrew P. McCormick. 1984. *Tuscany and the Low Countries: An Introduction to the Sources and an Inventory of Four Florentine Libraries* (Vol. 2). Amsterdam: John Benjamins Publishing Company.

Vignau-Wilberg, Thea and Hoefnagel, Jacob. 1994. *Archetypa Studiaque Patris Georgii Hoefnagelii, 1592: Natur, Dichtung und Wissenschaft in der Kunst um 1600: Nature, Poetry and Science in Art around 1600.* Berlin: Staatliche Graphische Sammlung.

Werrett, Simon. 2019. *Thrifty Science: Making the Most of Materials in the History of Experiment.* Chicago: University of Chicago Press.

Wheelock, Arthur K. 1999. *From Botany to Bouquets: Flowers in Northern Art.* Washington, DC: National Gallery of Art.

Wilkes, Benjamin. 1749. *English Moths and Butterflies.* London: Benjamin Wilkes.

Willughby, Francis. 1671. A letter of Francis Willoughby Esquire, of August 24, 1671. Containing some considerable observations about that kind of wasps, call'd Vespæ Ichneumones; especially their several ways of breeding, and among them, that odd way of laying their eggs in the bodies of caterpillars, &c. *Philosophical Transactions of the Royal Society of London 6*(76): 2279–281.

Worster, Donald. 1994. *Nature's Economy: A History of Ecological Ideas.* Cambridge: Cambridge University Press.

Index

The index does not reference subjects covered widely throughout the book, e.g. the greater part of Merian's biography and her primary research focus, the metamorphosis, behaviors and food choices of insects. The index does include more narrowly focused subtopics regarding Merian and her discoveries about insects.

Printed in the United States
by Baker & Taylor Publisher Services